The Quiet Limit of the World

The Quiet Limit of the World

Wayne Grady

A Journey to the North Pole
to Investigate Global Warming

Macfarlane Walter & Ross
Toronto

Macfarlane Walter & Ross
37 A Hazelton Avenue
Toronto, Canada M5R 2E3

Canadian Cataloguing in Publication Data

Grady, Wayne
    The quiet limit of the world: a journey to the North Pole
to investigate global warming

Includes index.
ISBN 1-55199-014-8

1. Global warming.   2. North Pole - Discovery and exploration.
I. Title

QC981.8.G56G73 1997   363.738'74   C97-931917-x

Printed and bound in Canada

Macfarlane Walter & Ross gratefully acknowledges the support of the Canada Council for the Arts and the Ontario Arts Council for its publishing program.

For Merilyn, *stella polaris*

And to the memories of Steve Hemphill
and Ikaksak Amagoalik

Contents

	Acknowledgments	ix
	Map: Arctic Ocean Section '94	xii
1	Water World	1
2	Taking the Ice	19
3	Killer Blue	37
4	The Lower Depths	53
5	*Fram's* Wake	69
6	Of Ships and Ice	83
7	And Bears	101
8	The Sea Beneath Us	113
9	Handfuls of Dust	127
10	Life on an Unknown Planet	141
11	Alphabet Soup	155
12	Dirty Ice	171
13	Losing Contact	187
14	Hydrometer Rising	201
15	Ultima Thule	219
16	The Channels	239
	Appendix	251
	Index	255

Acknowledgments

T HIS BOOK is the story of a scientific expedition to the North Pole, a voyage of exploration known to science as Arctic Ocean Section '94. It was a two-month, 7,000-kilometre trip involving 70 scientists and 250 Coast Guard officers and crew, so there are a lot of people to thank. First among them are Knut Aagaard, the overall expedition leader, and Eddy Carmack, the Canadian science leader, both of whom were more than generous with their time and friendship, not only during the trip but even more so over the long months of research and understanding that followed it. The American science leader, Art Grantz, was also extremely helpful. Among the many other scientists who spoke to me at length and with infectious enthusiasm, I especially want to thank Malcolm Ramsay, Robie Macdonald, Chris Measures, Mary Williams, and Lisa Clough. It was Caren Garrity who first mentioned me to Eddy as a candidate for the trip, and I am deeply indebted to her and her husband, René Ramseier, not only for their confidence but also for first sharing with me a portion of their limitless fascination with ice.

I would also like to thank the two icebreaker captains, Captain Phil Grandy (now retired) of the Canadian Coast Guard and Captain Lawson Brigham (also retired) of the U.S. Coast Guard. Both of them opened their minds and their ships (and, in Captain Brigham's case, his library) to me at critical points in this story. Mike Hemeon, chief officer on the *Louis*, was always ready with a thoughtful explanation, and Jim Purdy, the *Louis*'s logistics officer, was always handy with a key. While I was on the *Polar Sea*, executive officer Mike Powers lent me his office, an act of kindness for which I am very grateful.

Gary Ross's firm editorial guidance, both on the page and over evening drinks in Garyland, helped me and this book through several summer gales; and Jan Walter and John Macfarlane continue to know how to make a writer feel like an important part of the publishing process. I also thank Barbara Czarnecki for her patient and intelligent copy editing.

The generous and enduring friendship of Arnie Gelbart of Galafilm played an enormous role in the telling of this story. I also want to thank Stefan Nitoslawski for making his tapes and slides available to me during the writing. And thanks to Arnie, Tom Carpenter, and Andrew McLachlan, I was able to keep in touch with my wife, Merilyn, almost all the way to the North Pole; as always, her love, patience, and support, during both the long voyage and its longer aftermath, have made this book possible.

The title, a line from Tennyson's poem "Tithonus," was very generously lent to me by Ronald Wright.

As our two ships were parting company at the eastern edge of the polar ice pack at the end of AOS '94, Walt Olson, one of the technicians from the *Polar Sea*'s geology program, came up to me on the ice, poked his finger in my chest, and said: "Tell the whole story, man. No bullshit." It was good advice and I thank him for it. I hope I have taken it.

Here at the quiet limit of the world,
A white-hair'd shadow roaming like a dream
The ever-silent spaces of the East,
Far-folded mists, and gleaming halls of morn.

— TENNYSON, "TITHONUS"

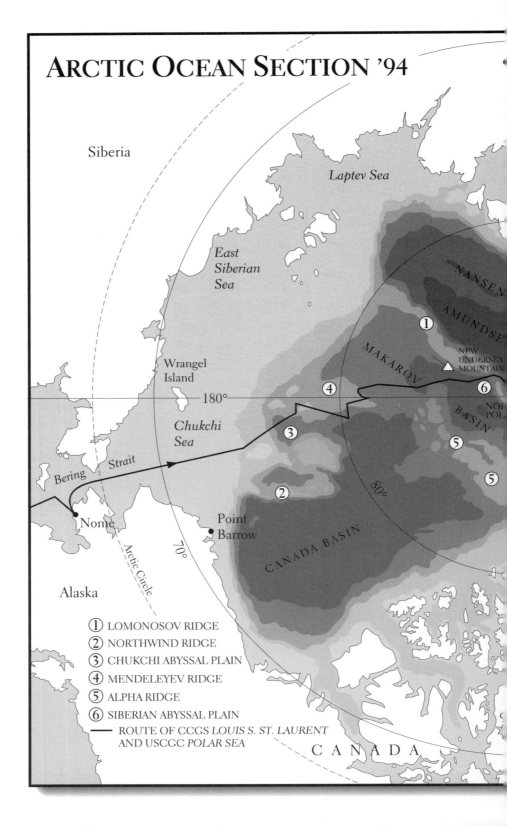

ARCTIC OCEAN SECTION '94

Siberia

Laptev Sea

East Siberian Sea

NANSEN

AMUNDSE

MAKAROV

①

NEW UNDERSEA MOUNTAIN

Wrangel Island

—180°

⑥

④

Chukchi Sea

③

NOF POL

BASIN

80°

Bering Strait

②

⑤

⑤

Nome

Point Barrow

CANADA BASIN

Arctic Circle

70°

Alaska

① LOMONOSOV RIDGE
② NORTHWIND RIDGE
③ CHUKCHI ABYSSAL PLAIN
④ MENDELEYEV RIDGE
⑤ ALPHA RIDGE
⑥ SIBERIAN ABYSSAL PLAIN
— ROUTE OF CCGS *LOUIS S. ST. LAURENT* AND USCGC *POLAR SEA*

C A N A D A

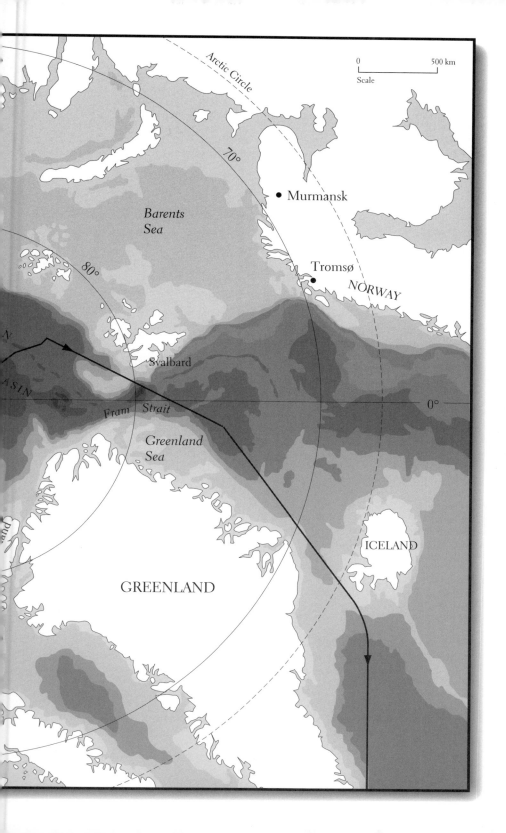

1

Water World

A departure is always good, or at least good enough.

— JOSEPH CONRAD, *The Mirror and the Sea*

I WAS HEADING SOUTH toward the American border when I drove into the rain. I'd been camping, hiking out with park wardens and wildlife biologists to look at coyotes for a book I was writing, and so had been out of touch for a couple of weeks. I hadn't heard about the rain. I spent the tail end of the day under my tent fly in Waterton Lake Provincial Park, in southern Alberta, watching a pair of magpies work their way across a sodden meadow beside the swollen Belly River.

In the morning, the gravel around the tent had been washed away and it was still raining. At the border, where Waterton becomes Glacier National Park and Alberta becomes Montana, the uniformed customs officer came out of her booth carrying an umbrella and a wet clipboard.

"Are you bringing any agricultural products into the United States?" she yelled at me.

"I don't think so," I yelled back.

"No citrus fruit? Oranges, lemons, grapefruit?"

"Well," I said, "I do have a few lemons in the cooler." Along with a bottle of vodka, a jar of Mott's Clamato Juice, and two cans of Campbell's Beef Bouillon. No one in the States, I knew, had heard of Bloody Caesars. "Do you want them?"

She looked at me. "No, thank you," she said. "Just don't peel them until you leave the country."

In northern Montana, it was raining so hard I decided to stop for lunch at a roadside restaurant rather than make sandwiches from the trunk of my car. There was a telephone booth outside the restaurant, and I tried to call home, but the rain was beating so loudly I couldn't even hear the dial tone. Inside, three truck drivers were eating club-house sandwiches and talking about the floods.

"What floods?" I asked, sliding into the next booth.

They looked at me. "Where the hell you been?" one of them asked.

"Canada," I said.

"Well, you sure ain't been to Kansas City. I just got across before they closed the bridge. They're putting up sandbags to keep the Missouri River from running down Independence Avenue."

"It doesn't seem to be raining that hard," I said.

"If you want to see hard," said the trucker, "you just keep driving south."

I drove south. In Choteau, Montana, the rain was falling in sheets. I stopped at another pay phone, this one bolted to the side of a sport shop that advertised 64 different kinds of gunpowder, and tried to call home again. I spoke briefly to my wife, who told me that the flooding in Missouri had been on the news. Water ran over the tops of my boots and trickled off the brim of my hat. I continued south. At the Freeze-out Lake Bird Sanctuary, I stopped to watch a large flock of American white pelicans circling high in a gray sky over a gently undulating prairie. Farther on, I saw an osprey diving for fish in a wide, flat river whose water had risen beyond its banks. Across the highway, a dozen common terns skimmed over a flooded field, their wings bending the tall blades of prairie grass poking up through the rain-dimpled surface of the water. All seabirds, I thought. I got back into my car. On July 9, I crossed the Missouri River. It didn't look out of control, but it did seem brown and vigorous for the time of year. This was supposed to be the dry season.

My business was with coyotes, so when I hit Route 68 I turned east and drove through Bozeman and then south into Wyoming. The rain dogged me. In Yellowstone, I set up my tent facing south and stretched the fly between it and a young pine, one of the few that had escaped the 1988 inferno that had turned three-quarters of the park into a forest of blackened spikes. At night I sat under the fly drinking Caesars and reading Ruth Rendell. During the day I hiked the Lamar Valley with a field researcher who knew where a pair of coyotes had maintained a den that spring. One afternoon, as we sat high up on the valley wall overlooking the Lamar River, the sound of a helicopter emerged from the roar of the rain. Kezha huddled under her rain gear and radioed the ranger station in Mammoth for news: a young fisherman, a 20-year-old boy, had had his hip-waders filled by the bloated river and was presumed drowned. All they found was a pair of sunglasses.

When my work was finished, I drove home through the North Dakota badlands and the Minnesota flats almost without stopping and never without the windshield wipers slapping time to the country music I had to listen to between newscasts. By the time I had crossed the last bridge over the Mississippi River that was still open to traffic, I had pieced together the story of the wettest summer of the century in the Midwest. And I had begun to appreciate what some people meant when they talked about serendipity and climate change, about how easy it is to tip a scale.

The Great Flood of 1993, as it came to be called, actually started in the autumn of 1992, when unusually high rainfall in the upper Mississippi Valley saturated the ground so thoroughly that any more water that fell on it instantly drained off into the river and its tributaries. This caused minor flooding in Iowa, Wisconsin, and Minnesota. An exceptional snowfall that winter prevented the ground from drying, and the spring runoff, combined with more rain that started in April, caused early flooding along the Missouri River. Fresh thunderstorms moved into Oklahoma in May, and by the end of that month an entire 260-kilometre stretch of the river, from Kansas City to Booneville, Missouri, was closed to commercial traffic by the U.S. Army Corps of Engineers, for fear the barges' wakes would damage the levees the corps had built along the riverbank. Still it rained. It rained in South Dakota and flooded the Big Sioux River; it rained in Minnesota and caused the Rock River to slip its banks; it rained so hard in Iowa — 15 centimetres of rain in an hour and a half — that the Des Moines and Raccoon Rivers doubled in volume and spilled over onto 200,000 hectares of prime farmland.

Then, when everyone in the nine Midwestern states was wondering just how much more rain they could take, it really began to rain.

In early June, an abnormally high pressure system over the Atlantic Ocean pushed a huge ball of warm, moist air from the Gulf of Mexico up the Mississippi Valley all the way to Minnesota. At the same time, a mass of Pacific air came rolling up from the southwest, hit the warm Gulf air over the Mississippi, then sank quickly under it and condensed into clusters of thunderstorms that dumped millions of litres of water on an already saturated land. Pickstown, South Dakota, which normally receives 29 centimetres of rain from April through July, received 64 centimetres, most of it in June. Waterloo, Iowa, got 77.5 centimetres in the same period, twice its normal amount. Farther downriver, in Davenport, Iowa, where the levees containing the Mississippi River are 4.5 metres high, the river rose 7 metres. In Des Moines, the flood stages are 4.5 metres high; the Des Moines River rose 10.5 metres. In Kansas City, on July 8, the Missouri River crested at 15 metres, 5 metres above the flood stages. By the first of August, the crest had reached St. Louis, where an estimated 28 million litres of water flowed past the Gateway Arch *every second*.

The damage to cities was swift and devastating — by mid-August, 50 people had been killed and 70,000 were homeless — but it was the farmers who were hardest hit. Desperate municipal workers in Prairie du Rocher, Illinois, used dynamite to blast breaches in the town's secondary levees to relieve the pressure that had built up on the main flood stages. They saved the town, but the resulting flood washed out 19,000 hectares of farmland. By the end of August, when the rain finally let up, 100 rivers had flooded their banks, overflowing 31 federal and 800 private levees and inundating more than 8 million hectares of farmland. Nine states — North and South Dakota, Iowa, Illinois, Kansas, Minnesota, Missouri, Nebraska, and Wisconsin —

were declared disaster areas. Of the $12 billion worth of damage inflicted on those states, $8 billion of it was sustained by farmers.

By the time I arrived home in Ontario, the worst of the flooding was over and water levels had begun to recede, although the rain kept up right through to September. A Gallup poll showed that 18 percent of Americans believed that, like the first Great Flood of the Bible, the 1993 flood was retribution from God for America's "sinful ways." The U.S. Meteorological Office offered a more mundane explanation. The high-pressure system that had pushed warm Gulf of Mexico air up the Mississippi Valley had been caused by elevated surface-water temperatures in the mid-Atlantic Ocean. And the blast of Pacific air from the southwest had been diverted in that unusual direction by an unscheduled El Niño event over the Humboldt Current, off the west coast of Chile. The Great Flood, in other words, was a local manifestation of global forces working on distant portions of the planet. A lot of people, including me, found this explanation even more unsettling than the prospect of being targeted by a spiteful, slate-cleaning God. We remembered what some scientists had been telling us for the past five years, that one of the signs of global warming was unusual and unpredictable weather. Whatever else it had been, the Great Flood of 1993 had certainly been that.

At home, I found a fax waiting for me from Caren Garrity, an ice scientist working for the federal Atmospheric Environment Service in Ottawa. Her specialty was the use of satellite imagery to predict global weather patterns, especially Arctic weather patterns and ice movement. I had met her in Gander, Newfoundland, in 1988 while she was participating in ice-related experiments in the Labrador Current. We struck up a friendship, and the next year I spent two months with her and her husband, who was also an ice scientist, in the Greenland Sea aboard the German research vessel *Polarstern*.

"A group of Arctic scientists are planning an expedition to the High Arctic next summer, to study the effects of global warming on the polar ice pack," her fax read. "They're looking for a writer to make the trip with them. How would you like to come along?"

I faxed her immediately: "Sorry I'm late — have I missed the boat?"

She faxed back that night. "You should fly to Sidney B.C. next week to meet with the chief scientist. His name is Eddy Carmack. You'll like him."

Eddy Carmack met me at the airport. A lean, young-looking man in his mid-40s with a West Coast tan and thick black hair, he was wearing khaki shorts and brown suede shoes. When he shook my hand his smile was warm and his voice soft, with an Arizona tang in it. He picked up the heaviest of my bags without the slightest attempt to pretend that it wasn't heavy and led me through the airport to the parking lot, where he heaved the bag through the side door of a bright yellow 1974 Volkswagen Westphalia camper van. He apologized for the van the way some people apologize for their dogs, saying he could have signed out one of the Institute of Ocean Sciences vehicles instead, but he'd been in a hurry. There were so many dents in the van it looked like an old hat. A green Coleman stove was bungeed to the top of a cabinet. The seat belts hung from bent coat hangers screwed to the doorposts, and the rear seat did not appear to be bolted to the floor. The engine purred tinnily, in the manner of vw engines, and neither of the side mirrors lined up with the road behind us. Eddy and his wife and their three kids did a lot of camping on the island, he said, and we talked about camping for a while.

"Where do you go?"

"Oh, we just head out, you know," he said. "Up the highway. There's only the one highway on the island. We like to go to Strathcona Park and do some camping."

"Did you get much rain this summer?" I asked.

"Not as much as some," he said.

We stopped at the Sidney Hotel, a low green building at the end of Sidney's main street overlooking a marina. I checked in, dumped my luggage in a long, narrow room with a view of the Strait of Georgia and, somewhat farther out, Mount Baker, and got back into the van. On the way to the institute, we talked in a roundabout way. Eddy's mind was a kind of inward extension of his van. It collected images, rearranged them into different patterns, and turned them into new ways of seeing old things. As we drove along the Saanich Highway, he pointed out landmarks sacred to the Saanich Indians — Mount Newton (a low, rounded mound with a radar dome perched on top like a giant golf ball), arbutus trees (which the Saanich will not cut down) — and told me that at the institute he had just hired a young Saanich student who was to spend the summer interviewing elders and making a map of the local coastline in Sencoten, the native language of the Saanich. "All these geographical features," he said, "the bays and coves and points and so on, had Sencoten names long before we got here, and those names were important not only historically, but also because they made sense from a scientific standpoint. They would describe some physical peculiarity of the spot — a good place to hunt, for example, or a bad place to launch a canoe, that sort of thing. Mount Newton's Sencoten name means place of refuge: they didn't name it after some defunct farmer who came from England in 1851, like we did. They named it with something they could use."

Driving over a dried-up creek bed made him think of an environmental hydrogeologist he'd recently invited to give a talk at the institute on stream restoration. "He showed us how rivers are structurally like trees," Eddy said, "if you look at them from the air. They both have a main trunk and 3.6 large branches, each of which has 3.6 smaller branches, and so on. Both the tree trunk and the river pattern are ideally suited to getting water from point A to point B in the most efficient manner possible. If the landscape doesn't allow a river to have 3.6 branches, then the river changes the landscape until it does. If we change the landscape, then the river suffers, the same as a tree suffers when we lop off a major limb. It was an interesting talk," he said. "It made me realize that everything is interconnected in ways we haven't even begun to understand."

Eddy is a physical oceanographer. His first degree, from Arizona State, was in math, and he went on to spend six years studying oceanography at the University of Washington. "I figured a career in oceanography was a good way to get out of Arizona," he said. He studied under Larry Coachman, known as Coach to his students and one of the best high-latitude oceanographers in North America, and then under Knut Aagaard, one of Coach's former students. "Knut wasn't much older than I was," Eddy said, pronouncing the K in Knut so that it sounded like Canute. "We met in an ice camp on the Beaufort Sea in the late 1960s; we were taking part in a study of the effects of wind stress on sea ice, and spent most of the spring playing chess in a little plywood shed with a hole in the floor." Aagaard had just returned from completing his postdoctoral work on deep-water formation in the Greenland Sea at Norway's Geophysical Institute and was going back to Washington in the fall to teach. Eddy was his first graduate student. "I don't think either of us knew what to expect of the other," Eddy said, "but we became pretty good friends and I certainly learned a lot."

After getting his PhD, also on deep-water activity in the Greenland Sea, Eddy and his wife, Carole, who had been a librarian in the university's oceanography library, moved to La Jolla, California, to work at the Scripps Institution of Oceanography. Eddy wanted to combine his new interest in deep-sea diving with a study of Antarctic ocean currents. But when Carole was offered a job at the Vancouver Public Library, the couple moved back to the Northwest. For a while, Eddy worked as a limnologist, studying lakes in the British Columbia interior with Environment Canada, before transferring to the Institute of Ocean Sciences in Sidney in 1986.

The IOS, along with the Bedford Institute of Oceanography (BIO) in Dartmouth, Nova Scotia, constituted the oceans half of the Canadian Department of Fisheries and Oceans. Most DFO scientists spend their time on the fisheries side, trying to find things that went wrong with the cod stocks or spawning salmon that are not embarrassing to the government. A lot of fishermen and not a few journalists were fond of reminding the DFO that it had been studying cod and salmon populations for 20 years and yet had not been able to foresee, let alone forestall, the sudden disappearance of the B.C. and Newfoundland fisheries. The scientists at the IOS and the BIO knew all along that there was no way of predicting the future until they had a firm grasp on the past, and they are a long way from that. "We still know very little about the Atlantic and Pacific Oceans," Eddy said. "And we know less than that about the Arctic."

When the IOS was formed, in 1977, it was housed in an old sea-plane hangar and a few outbuildings near the Victoria airport, which is in Sidney. Now it has its own complex, a group of brand-new, tinted-windowed brick buildings on a hillside overlooking the federal dockyard, where DFO inspection boats tie up beside fishing trawlers and brick-colored pigeons compete with herring gulls for scraps of fish. Inside were the usual fluorescent light fixtures, freestanding room dividers, and Xerox machines, but Eddy's office, at the end of a large common area filled with chart tables, was a lot like his van — an eclectic collection of items subtly related to one another by his own personality. Entering it was like stepping into a fertile mind. His file drawers, for example, were identified not by the usual labels but by baseball cards and card-sized photographs of Arctic explorers: Roald Amundsen, Joseph-Elzéar Bernier, Gary Carter, Joe DiMaggio. The effect was pleasantly dizzying. Some file boxes were labeled twice, once with the arcane nomenclature of ocean science and once with equally arcane baseball cards; sometimes the conjunctions made sense and sometimes, at least to me, they didn't. Biophysics had a Tom Henke card, and I could understand that. Chaos had George Bell, okay. But Halothermodynamics had Dale Murphy, Double Diffusion had Kirby Puckett, and Lake Climate had Willie McGee. Above the desk, a large chrome ring dangled from the ceiling bearing a brightly painted wooden parrot with a packet of crackers taped to its feet. On the desk itself, shuffled into a stack of official-looking papers, was a child's drawing of a tall man flying a red kite.

Leaning back in his swivel chair, Eddy seemed perfectly at ease among these disjuncta. I imagined that all the normal office stuff — stapler, pencil sharpener, pens — was in his house, on the dining room table. Stress therapists call this contiguity: no psychological breaks between real life and work life.

Eddy's great idea — to gather together 70 of the best ocean scientists in North America, put them on two large icebreakers, and take them to the North Pole — came to him in the form of a joke during a meeting of the U.S. Polar Research Board in 1989. The board had listened to a report from a European group that intended to make an expedition on the *Polarstern* and the Swedish icebreaker *Oden* to the

North Pole, going from Tromsø, Norway, north through the Greenland Sea. Eddy had leaned over to Knut and whispered: "You call your Coast Guard and I'll call mine, and we'll come up the other side and meet them at the North Pole."

Knut had laughed. No ship had ever made it to the North Pole from the west. No ship had even tried for the past 80 years. More than 70 percent of the polar ice cap sits in the western side of the Arctic Ocean, like a lop-sided beanie, between the North Pole and Bering Strait. The whole area, some 200 million square kilometres of ice, was unexplored. That night, however, over drinks in the bar, the subject came up again.

The word "unexplored" does something to a scientist's mind. So much of science consists of testing someone else's hypotheses, duplicating someone else's experiments, gathering someone else's data. The western Arctic Ocean was to Eddy and Knut what central Africa had been to Livingstone, what the American West had been to Thomas Say; the icebreakers would be to them what the *Beagle* had been to Charles Darwin. As Eddy and Knut talked it over, they thought they might be able to make their respective governments see some political advantage in the scheme: "We thought it was really important," Eddy said, "in terms of both countries' national interests, that a statement be made about our capabilities and influence in the High Arctic."

But it was Knut who came up with an even more compelling argument. "The purpose of this expedition," he said, "should be to study the Arctic in the context of global warming."

After that meeting, Eddy returned to Sidney and put a call through to Captain David Johns, then chief of northern operations for the Canadian Coast Guard. Eddy had never met Captain Johns before and wondered whether he was doing something rash. "Put yourself in my place," he said. "I mean, I just picked up the phone, asked to speak to the director of northern ops, and said: 'Can I have your largest and most powerful icebreaker for a trip to the North Pole, please?' I felt pretty foolish."

"You said North Pole?"

"Yes," Eddy replied. "At that point, we knew we were talking about an ocean crossing, and if you're going to cross the Arctic Ocean you might as well cross it over the Pole. Anyway, the first thing he asked me was whether there was a scientific reason for going to the Pole, and I said there was an excellent scientific reason, even if you don't count global warming research. The Canadian Basin, the section of the Arctic Ocean between the Pole and Canada, is the last ocean basin in the world that has never been crossed by a modern research vessel. It is a huge unknown if you're trying to put together a picture of ocean circulation, or distribution of marine biota, or the movement of certain chemicals and nutrients and oxygen. The Canadian Arctic Ocean is the last missing piece of the puzzle. It's our own backyard, and we know practically nothing about it."

It was also a propitious year to be making proposals for an investigation into global warming. Before the Reagan administration, at the behest of the oil and coal cartels, put a freeze on funding for global warming research, almost every major science center in the world was issuing warnings about the long-term effects of a measurable increase in the average temperature of the earth. The decade of the 1980s had been declared the hottest ten-year period since we began keeping records. The hottest years, in descending order, were 1988, 1987, 1983, 1981, 1980, and 1986, and, although no one knew it then, 1990 would be hotter than 1988. (Currently the hottest years in history have been, in descending order, 1996, 1995, 1994, 1993, and 1992.) In 1989, sweltering heat had prostrated the citizens of New York City, the temperature in Washington, D.C., soared to a record 38.3 degrees Celsius (100.9 degrees Fahrenheit), drought devastated crops in the Midwest and California, and — ominously, in light of what happened in the summer of 1993 — water in the Mississippi, Missouri, and Ohio Rivers was the lowest it had been since 1872.

Although all these indicators could be dismissed as caprices of weather, they emphasized what scientists had in mind when they talked about global warming. If we're right, the scientists could say, it'll be like this all the time from now on, except worse. No scientist would say that before 1988. At the first meeting of the Senate Committee on

Energy and Natural Resources, held in Washington in June 1988, James Hansen of the NASA-Goddard Space Institute was the first scientist on record to come out and state unequivocally that the earth's average temperature was rising. The chances, he said, that the observed changes in weather patterns worldwide were natural or normal or temporary were about one in 100: "It is time to stop waffling so much and say that the evidence is pretty strong that the greenhouse effect is here."

At a second Senate committee hearing, in August 1988, physicist Stephen H. Schneider of the National Center for Atmospheric Research was asked the obvious question. The average temperature in North America is 1 Celsius degree higher than it was 30 years ago. A hot day in Dallas or Toronto is 31 degrees C instead of 30. So what? The question was outside the area of a physicist's expertise, and so far, apart from James Hansen, physicists had declined to answer it. In Canada, a similar question was posed to the House of Commons Standing Committee on the Environment in 1989. The committee found during its cross-country Forum on Global Climate Change that three out of ten Canadians believed global warming would actually be good for Canada: "If there are going to be warmer winters," the committee was told, "we are going to be able to grow more food." Would it be a bad thing, the committee was asked, if the Corn Belt shifted north from Kansas to Saskatchewan, or if tankers were able to get into Churchill, Manitoba, year-round, or if the growing season in the boreal forests of northern Ontario were extended by a month or two? Would it spoil some cosmic plan, Schneider was asked in Washington, if the growing season in Kentucky were extended another week on either side?

Well, yes, said Schneider, it would. An average rise in temperature of 1 Celsius degree (or 3 Fahrenheit degrees) in, say, Washington, tripled chances of having five days in a row of 35-degree-C (95-degree-F) heat in July, from one in six to one in two. In Dallas, the odds of having five 38-degree-C (100-degree-F) days in a row jumped from one in three to two in three. Schneider pointed out that extended heat waves would have serious health consequences for elderly and asthmatic citizens, the cost of air-conditioning would double, as would the amount of CFCs released into the atmosphere, and more power plants would have to be built. Moisture loss in crop-lands would increase, not only lowering yields and placing stronger reliance on (gas-powered) irrigation systems, but also increasing the amount of airborne water vapor — a natural but potent greenhouse gas — which would in turn accelerate the greenhouse effect.

Scientists in Canada told the Environment Committee that "global warming is likely to have significant negative effects in Canada, as elsewhere in the world." One of these negative effects would be the elimination of the entire boreal forest west of James Bay, which is already under severe climate stress; another would be a significant drop in the water level of the Great Lakes. A 1989 study by the Land Evaluation Group at Guelph University predicted extensive crop fail-ures in Ontario's agricultural heartland, especially for corn and soy-beans. Increased moisture stress from higher evaporation and rainfall would more than offset a longer growing season — the crops would just have more time in which to rot. The cost to Ontario's agricultural community would be $170 million a year. A similar study conducted in Saskatchewan suggested that, if average summer temperatures in-creased by only 2 Celsius degrees, the wheat-growing region of Canada would experience prolonged droughts, similar to those of the Dirty Thirties — which cost the country $1.8 billion in lost revenues — only more frequent, and more acute. And all these predictions were based on a temperature rise of 2 Celsius degrees by 2030: no one could say how high the average would go if current trends continued; no one could say it would stop at 2 degrees or 5 degrees, or if it would stop at all.

Both the U.S. Senate committee and the Canadian Forum on Global Climate Change were told over and again that more informa-tion was needed before an accurate assessment of the effects of global warming could be made; more scientists needed more money to go more places to gather more data. Scientists are always saying that, of course, but this time they could be specific about where they needed to go and what they had to look for. Computer-model predictions of climate change were being run with little input from the oceans. The massive World Ocean Circulation Experiment (WOCE), inaugurated in 1990 to piece together a global map of the world's oceans and their interdependencies, did not call for a single look at the Arctic Ocean. No one knew what was happening up there. Oceans in general are notoriously hard to quantify in computer models: two computer models using slightly different guesses about how a change in ocean circulation would affect global warming could come up with completely different scenarios. But the Arctic Ocean was a complete mystery. Schneider's book *Global Warming: Are We Entering the Greenhouse Century?* (1989) didn't once mention the Arctic; neither did John Firor's *The Changing Atmosphere* (1990); neither did Thomas Levenson's *Ice Time: Climate, Science and Life on Earth* (1989), an otherwise exhaustive account of the changing climate and our attempts to understand it. Fluid dynamics, Levenson pointed out, the branch of physics concerned with the interaction between the oceans and the atmosphere, "is fiendishly complicated"; Werner Heisenberg, the inventor of quantum physics, had tried his hand at it and given it up because there were too many variables and not enough data. Albert Einstein found fluid dynamics "too difficult."

In the case of the Arctic Ocean, there were virtually no data at all. No one knew whether it acted as a heat sink or the world's largest refrigerator. Was it absorbing greenhouse gases from the air, or did it release them from its surface water? Was it speeding up the greenhouse effect, or was it our last bulwark against mass desertification? What was going on under all that ice? No one knew. Someone, it was strongly suggested, should go up there and find out.

That's about when Captain Johns got Eddy Carmack's phone call.

"He was interested right from the start," Eddy says. "All we had to do was come up with a plan, a budget, some money, and 70 scientists. And all he had to do was find us a ship big enough to take us to the North Pole."

It took five years. Funding for global research still wasn't thick on the ground. Even so, when Knut and Eddy sent out calls to see who wanted to go to the Arctic with them, they received enough program proposals to fill four ships. It took them years to winnow the list down to 70. Now they were looking for a writer, a filmmaker, and two artists. "It's traditional to take a nonscience party along on an Arctic expedition,"

Eddy said. "At least in so far as there is an Arctic expedition tradition. Are you interested? We're leaving from Victoria on July 17."

"To realize Victoria," Rudyard Kipling remarked in 1907, "you must take all that the eye admires most in Bournemouth, Torquay, the Isle of Wight, the Happy Valley of Hong Kong, the Doon, Sorrento, and Camps Bay, add reminiscences of the Thousand Islands, and arrange the whole around the Bay of Naples, with some Himalayas for the background."

Kipling was talking to graduate students at McGill University in Montreal, where he had just received his first honorary degree, and it's a safe bet that few of his listeners had a clear mental picture of Torquay or the Doon or, for that matter, of Victoria, by which to judge the accuracy of Kipling's description. I know I certainly didn't when I arrived, ready to embark on the expedition. It was a scorching July day, and the narrow streets were awash with taxis and buses, tourists and buskers, teenagers on inline skates, teachers in Winnebagos and VW vans, hippies in ponchos, and bankers in sunglasses and T-shirts with stylized Haida salmon on their chests. Everyone seemed to be either retired or on vacation, strolling around as though they had all the time in the world and nothing to do but shop. Every store sold souvenirs called "Genuine Native Handcrafts." Every corner was an art gallery specializing in the works of "Emily Carr and other local artists." The buildings that housed them were not high enough to block out the sun but had enough glass to act as solar collectors and enough stone to absorb and hold the heat, so that the streets between them seemed like microwaved gullies. It was as though everyone in Ottawa had suddenly moved to Banff, and taken their buildings with them.

I trudged along Government Street carrying my bulging leather book bag, my camera bag, a small suitcase, and my father's old navy kit bag, sweating like a grilled sausage through fork holes. Suddenly I emerged onto a broad, parapeted walkway that curved gently around the rim of the Inner Harbour. It was like a last-minute pardon from being roasted on a spit. A man in a tartan kilt and playing a bagpipe had attracted a bevy of elderly Scots tourists at the near end of the curve, and I stepped carefully through them before stopping to take in the view. Behind me, keeping a proprietary eye on the harbor, rose the stern stone splendor of the Empress Hotel.

When I turned my back on the Empress, the Inner Harbour stretched out lazily before me like an unfocused opalescent eye. Coming from Ontario, I was used to thinking of bodies of water as finite, contained; this harbor lay at the foot of the wide sea like a welcome mat.

My destination, the Laurel Point Hotel, was across from where I was standing. Water taxis were scooting like Jesus bugs back and forth across the harbor, and there was a stop on a small jetty directly below my feet. I hoisted my gear, descended the stone staircase, and boarded one of the corklike canopied boats. Within seconds we were in the middle of the harbor, heading in a roundabout way to the hotel. Seaplanes and helicopters roared overhead, and an immense blue-and-white ferry was being tugged gently toward its berth. As we crossed the ferry's wake and jaunted toward Laurel Point, I caught a glimpse of the larger harbor farther out, with its unclouded sky and promise of open water. The *Louis S. St. Laurent* and the *Polar Sea* were at anchor out there somewhere, waiting to be brought in to Ogden Point, and within 24 hours I'd be aboard the *Louis*, heading for Nome, Alaska, to pick up the scientists. From there we would sail north until we reached the North Pole. I remembered the rest of Kipling's speech, in which he rejected the notion that Victoria was a pale imitation of England. "For no England," he said, "is set in any such seas, or is so fully charged with the mystery of the larger ocean beyond."

After checking into the hotel, I bought a raft of newspapers and installed myself in the lounge. The papers were once again full of accounts of flooding, this time in the lower Mississippi Valley in the wake of Tropical Storm Alberto, which had hit Florida on July 3 and moved north. Thirty people had been killed and 50,000 evacuated from their homes. The floods of the previous year in the Great Plains had finally gravitated to the south. The rain had lasted only ten days, but that was long enough to put Florida, Georgia, and Alabama into a state of emergency. In Albany, Georgia, the Flint River peaked at 14.6 metres on July 11, 8.5 metres above the flood stage: 295 billion litres of water flowed over the levees and onto prime cropland in a single day. Farther north, Americus, Georgia, received 54 centimetres of rain in 24 hours. Sixteen people had drowned. It was the worst flooding the region had experienced since 1920.

In the entertainment section, I read that Kevin Costner was raising money for a new film, a futuristic fantasy based on the idea that the world's oceans would one day reclaim the planet and everyone would live on floating mechanical islands. The film would be called *Waterworld*. "Where," asked one film columnist, "does Hollywood get its ideas?"

2

Taking the Ice

Whenever I enjoy anything, I always look forward to writing it down.

> – CHARLES DARWIN, in a letter written aboard
> the *Beagle*

W E ARRIVED at Nome, Alaska, at 10 o'clock on the morning of July 24, a week after leaving Victoria. The temperature in Victoria had been 30 degrees C; in Nome it was 8.6. But the sun was shining and the air was crisp and clear. According to the *Marine Pilot*, which I read on the *Louis's* bridge, we were fortunate: Nome gets only five cloudless days from June to October, and this seemed to be one of them.

The harbor, as we sailed into it, was awash with swirls of gray-backed gulls — adult glaucous gulls (*Larus hyperboreus*), as nearly as I could determine from my Peterson's guide. There were a few different-colored gulls wheeling about as well, most of them varying only slightly on the theme of white bodies, gray wingbacks, and black primaries. Gulls all look alike to me. The adults are bad enough, and the juveniles are impossible to distinguish, most of them bearing little resemblance to what they will look like as adults, a stage of maturity it takes some of them three or four years to achieve. In the meantime, they go through speckled, mottled, blotched, and piebald phases, each species in its own highly individualized way. Riding on the water beside the ship, for example, was a juvenile gull that was mostly white, but with a strip of brown acne across its shoulders that meant, according to Peterson — who devotes three pages to immature gulls — that it could have been either a first-winter western gull (*L. occidentalis*), a second-winter glaucous gull, or a third-winter herring gull (*L. argentatus*), take your pick. I picked glaucous.

"Back home," said Bing, a crew member who was looking down at the bobbing gull, "we call the older ones saddlebacks."

"Where's home?" I asked him.

"Halifax," he said, and we both leaned on the rail, me wondering how a word like "saddleback" had crept into a maritime vocabulary, Bing looking up at the line of rounded hills above Nome. "I guess this'll be my last kick at the can before I retire," he said, evidently feeling a bit saddlebacked himself. His hair, curling out from under his *Louis S. St. Laurent* ball cap, was iron-gray at the temples. "When I was told I could sign on for this trip, I didn't know what to say. I talked it over with my wife, and she said, 'You've got to go, you know,' which surprised the hell out of me. I said, 'You know I'll be gone for three months,' and she said, 'I know, but you've got to go for it. Everybody's talking about this trip, everyone wants to know about it, and you've got to go,' she said. 'It's an honor to be asked.' And I had to admit that that was exactly how I felt about it, too. It was an honor to be asked. So here I am."

Another crew member came along the passageway and stopped to look over the rail with us. He was wearing western boots, tight blue jeans with a huge belt buckle, and a T-shirt depicting a pack of wolves howling up at a cold winter moon. He asked to borrow my binoculars and, without removing his sunglasses, swept the almost featureless coastline with them.

"Jesus," he said quietly. "Jesus, that's beautiful, innit?" Bing and I nodded. "And me buddies back home asked me why I wanted to come on this trip."

Nome Harbor (as the slight indentation along the Seward Peninsula is called) was too shallow to admit seagoing vessels, so we anchored a mile offshore and lowered one of our Liberty boats over the side. Then we tried to get into it. At first we tried dropping the gangway and jumping in from the bottom platform, but the sea was so rough the last step swung up more than 3 metres above the boat, then dipped down nearly to water level, then swung back up. We had to time our jumps perfectly at the nadir of each swing; still, six of us, including Stefan Nitoslawski, a filmmaker, and me, managed to leap into the boat, cameras and tripod flapping, with no loss of life. At that point, Tom Lafford, the third mate, decided it would be safer to take the boat around to the leeward side of the ship and board it by means of the Jacob's ladder — two long ropes with wooden rungs between them that dangled precipitously over the rail and down into the boat. After a few harrowing adventures, 23 of us, our bright orange life jackets tied tightly around our throats, roared off toward the distant shore accompanied by the delighted laughter of half a dozen (possibly) glaucous gulls.

As we neared the dockyard, the variously colored shapes we had seen from the ship revealed themselves to be pastel-painted houses above rows of huge oil tanks and bobbing fishing boats. Stacks of white and red containers also lined the shore, with names of distant cities stenciled on them — Juneau, Seattle, Hong Kong, Rotterdam — suggesting that Nome is part of a vast international chain of exotic ports of call. Some of the stacks had large yellow earth-moving vehicles resting on top of them, like Dinky toys. We cut throttle and glided up to a small floating dock. Tom told us to keep our life jackets with us so they wouldn't be stolen.

"We'll meet back here in one hour," he said.

It was Sunday, and Nome seemed weirdly deserted. There was no one on the dock, no curious residents, no customs clerks or immigration officials. The muddy streets, as we fanned out into them, showed tire tracks, but no cars or trucks appeared to have made them.

"Set phasers to stun," someone said.

Looking down one empty laneway to the harbor, we could see the *Louis* and the *Polar Sea* resting calmly on a flat sea, their "at anchor" flags hanging limply in the still air.

Nome was founded in 1899, a year after rich deposits of gold dust were discovered in nearby Anvil Creek — its first name was Anvil

City, but that was soon changed to Nome, after Cape Nome, the promontory we had passed at the mouth of the harbor (the word is a misreading of the inscription "?Name" on Cook's 1778 chart of the coastline). By 1900 it was the largest city in Alaska, an instant metropolis with hotels, banks, two newspapers, and 20,000 citizens, most of whom had scurried from Dawson City at the first rumor of gold. Claims were staked along 20 kilometres of beach, practically the whole shoreline visible from our ships, and placer mines were worked all day. In 1903, $2.5 million worth of gold was shipped out of Nome; by 1920, the gold rush had rushed elsewhere, and Nome was a near ghost town of 982 people and a lot of empty gambling houses. In 1960, as part of the government's centralization plan, 2,000 Inuit were relocated to the town, and the population is now around 3,500. In the early 1970s, during the debate over moving the capital out of Juneau (where it had been since 1900), cases were made for relocating it to Fairbanks, Anchorage, Seward, Sitka (the old Russian capital), Fort Yukon, and even Kotzebue. For a while, Solomon-like, the government considered building an entirely new capital inland, near Mount McKinley, a kind of northern Brasília. No case was ever made for moving it to Nome. In the end, it was left in Juneau.

There were still three or four hotels in Nome, now called "tourist inns," all at the intersection of two unpaved streets at the waterfront, and all open for business. On the sidewalk beside the Nugget Inn was a plaque dedicated to the memory of Louise Forsythe, "the first girl to come to Nome, at age 12, on June 21, 1899." Louise married Michael Joseph Walsh on August 1, 1909, and the plaque had been placed in its stone pedestal in 1972 by their ten children. After reading the inscription, I felt an irresistible urge to phone my wife. In the Nugget, I joined a lineup in the lobby for the telephone. When my turn came, I left a plaintive message on the answering machine and went back out into the sunlight. Incredibly, since no roads lead to Nome, a huge tour bus had pulled up and disgorged a string of Japanese tourists, all of whom were filing nonstop into a gift shop whose door was across the sidewalk from the bus. I followed them in, carrying my day pack and my fluorescent life jacket, and eased delicately among glass shelves of exquisite Alaskan jade figurines of polar bears and caribou until I came to a rack of books. There were the usual tourist publications and a few good books, such as John Haines's *The Stars, the Snow, the Fire* and John McPhee's *Coming Into the Country*, which I bought. I was going to have a lot of reading time over the next two months.

Emerging into the hazy street, I noticed a man wearing a black bomber jacket with gold piping and a greasy baseball cap, sitting on the windowsill of the Nugget Inn, drinking coffee from a Styrofoam cup and smoking a cigarette. When he saw me he stood up and walked stiffly in my direction. His long black hair swayed under his cap, and his face was etched by more than the usual ravagements of a lifetime in the North. He asked if I wanted him to show me around town, and I told him I had to be back at the Liberty boat in 20 minutes.

"It won't take that long," he said.

First he led me around to the back of the hotel. Two mud-splattered all-terrain vehicles were parked outside an open door, through which I could see two men sitting at a table, drinking from cups and talking. Behind the hotel, on the seawall, silhouetted darkly against the glistening harbor, was a stack of antlers taller than my head, all weathered gray as driftwood. Some had skulls attached to them.

"Moose," said my guide.

We walked back around to the side of the hotel and he pointed to a sign I hadn't noticed before — a large wooden billboard flanked by two life-sized wooden lumberjacks and the words "Welcome to Nome, Alaska" carved into it, along with the information that the seawall had been constructed from 136,000 tons of granite blasted out of Cape Nome. It was an ornate sign. The edges had been scrolled into curlicues and arabesques, the letters were deeply gouged, and the lumberjacks had a certain rough-hewn handsomeness.

"Carved with chain saws," said my guide.

We walked along the sidewalk and stopped in front of the laundromat. In the tall grass beside the building lay what looked at first like a large, oddly shaped hunk of concrete, gray and pitted, with a round hollow core and two large wings spreading from the central hub.

"Whale vertebra," he said. "Been layin' there for years. Somebody found it down on the beach."

"Are you from Nome?" I asked him.

"Came here in 1960," he said. "From up the coast. Been here ever since, except for a year in Vietnam."

"What did you do in Vietnam?"

"Dental records," he said.

The last thing he showed me was a bust of Roald Amundsen, the Norwegian explorer, perched atop a large boulder in front of what looked like a community center. Amundsen's craggy, mustachioed face looked stern and astonished, as though he had just realized that he was embedded in solid rock. Bolted to the stone where his chest would have been was a bronze plaque:

Roald Amundsen (1872-1928)
Completed the first trans-polar flight
over the North Pole,
Europe – North America.
Left Spitsbergen May 11, 1926,
in the airship "Norge".
Landed at Teller, near Nome,
May 14th, after travelling 5,400 kms.

The monument had been erected in 1976, and an identical one was placed in Ny-aalesund, Spitsbergen, Amundsen's departure point.

Roald Amundsen shortly
before his death in 1928

"Teller's just up the coast a ways," said my guide.

I thanked him and shook his hand. He told me the quickest way back to the boat was to walk along the seawall, which I did, passing dilapidated houses and piles of truck parts on my right, and stacked containers and a small fleet of fishing boats bobbing against their tethers on my left. I arrived at the Liberty boat just as Tom was giving the order to cast off.

Although Amundsen is best known as the explorer who first sailed through the Northwest Passage, in 1903, and, eight years later, beat

Frederick Scott to the South Pole, he was also the first person to set eyes on the North Pole, a feat he accomplished in a dirigible at a time when it looked as though dirigibles and not airplanes were to be the long-distance flight mode of the future. Airships didn't burn vast quantities of fossil fuel, and they could carry more passengers than airplanes. They were also quiet and as luxurious as ocean liners.

Amundsen purchased the *Norge* from the Italian government, which had been using it (called something else) as a troop carrier. He paid about $10,000 U.S., an amazingly low price even in those days, but there was a catch: the man who arranged the sale, Umberto Nobile, insisted on coming along as Amundsen's pilot and bringing a crew of five Italians with him. Amundsen almost turned the deal down, since he didn't like Nobile and wanted the expedition to be a completely Norwegian venture. In the end, though, he gave in, because he believed in the future of lighter-than-air travel, and he wanted to be one of its earliest advocates. Flying a dirigible nonstop across the Arctic Ocean would be a fitting launch to a new era of silent, inexpensive transoceanic flight. Norway wasn't all that interested in funding the venture, and Amundsen had to raise the money in the United States, most of it coming from the adventurer Lincoln Ellsworth, who also wanted to come along.

Amundsen, Ellsworth, Nobile, and the *Norge* arrived in Spitsbergen on May 7, 1926, and began preparations for the flight. A few days later, Admiral Richard E. Byrd of the U.S. Navy arrived with his airplane, the Fokker trimotor *Josephine Ford*: Byrd was going to fly to the North Pole. Nobile was furious — he saw Byrd as a competitor for the glory that was supposed to be going to Italy — but Amundsen was more philosophical. Or perhaps shrewder: after all, what better way to herald the dawn of dirigible flight than to demonstrate its superiority over the internal combustion engine in the High Arctic? To Nobile's disgust and Ellsworth's amusement, Amundsen even helped Byrd get off the ground by lending him a pair of skis that could be modified for use as runners on the *Josephine Ford*. He refused to regard Byrd as a threat because, he said, Byrd was only going to fly to the Pole and then return to Spitsbergen. Amundsen wanted to become the first person to fly across the Arctic Ocean. "The Pole was not our objective," he wrote in *My Life As an Explorer*. "Our objective was Alaska, and the Pole was only an interesting incident."

Byrd took off at 12:45 P.M. on May 9 and returned the same day at 4:07, claiming to have made it to the Pole. Amundsen greeted him on the landing strip and congratulated the American warmly; by the time Amundsen's autobiography was published in 1927, Byrd's claim to have made it as far as the Pole was being seriously doubted, if not discredited, but Amundsen wrote highly of Byrd and made no reference at all to Byrd's detractors. He reserved his vitriol for Nobile.

The *Norge* lifted off from Spitsbergen at 10 in the morning of May 11 and headed boldly north with Nobile at the helm. It soon became apparent that Nobile, though he had helped design the dirigible, knew nothing about flying it. Worse, he was not good at keeping his head in critical situations. He couldn't point and steer at the same time. In fact, he couldn't stand still and steer at the same time. If he looked down toward the ice, the dirigible inevitably tilted dangerously downward. Instead of pulling back on the joystick, Nobile would let go of everything and run around the bridge shouting, "We're going to crash, we're going to crash!" Amundsen was too much the imperturbable leader to interfere with Nobile's position as pilot; more than once, however, Ellsworth, not bound by such Nordic niceties, ran over to the wheel and saved the ship from crashing into the ice. That they made it out of sight of Spitsbergen was a wonder; that they made it all the way to Teller was a miracle: 5,400 kilometres in 42 hours.

They did, apparently, fly directly over the North Pole, although Amundsen admitted that taking coordinates from a moving gondola 90 metres above a drifting ice cap was not a reliable method of determining position. He barely mentioned this astonishing bit of madcap navigation in his account: either he wasn't sure of his bearings or he was reluctant to admit that he hadn't seen any of the American flags that Byrd claimed to have dropped a few days earlier (Byrd later claimed he had "forgotten" to drop them).

But one of Amundsen's most scathing descriptions was of Nobile's behavior at the North Pole, and so he must have been aware of where he was. Amundsen and Ellsworth had brought along two small flags, one Norwegian and one American, each "not much bigger than a pocket handkerchief," which they planned to drop discreetly over the side as they cruised above the Pole. Why not? They were, after all, the first human beings ever to have set eyes on the spot, although they didn't know that at the time. "As we crossed the Pole," Amundsen writes, "we threw these overboard and gave a cheer for our countries. Imagine our astonishment to see Nobile dropping overside not one, but armfuls, of flags. For a few moments the *Norge* looked like a circus wagon of the skies, with great banners of every shape and hue fluttering down around her. Nobile produced one really huge Italian flag. It was so large he had difficulty getting it out of the cabin window."

After the *Norge* landed in Teller, Amundsen and Ellsworth left Nobile in charge of the airship while they walked the dozen or so kilometres to Nome to communicate their success to the Norwegian and American authorities. Imagine their astonishment again when, on their second day in Nome, they saw Nobile arrive like a dignitary aboard the U.S. Coast Guard cutter *Bear* and wire every major newspaper in the world with his own account of his own extraordinary feat, while Amundsen and Ellsworth continued to go through proper channels by notifying their respective governments first. Amundsen and Nobile conceived an implacable and rather public hatred for each other. But, two years later, when Nobile was lost in the Arctic after the crash of another airship, the *Italia*, Amundsen lost his life in the search for him. As Amundsen wrote in his autobiography, "whenever a group of men get together to do a difficult task, misunderstandings are bound to arise, clashes of temperament occur, incidents happen that are best forgotten. Polar expeditions are no exceptions to that rule."

I have never really understood how an exception can prove a rule, but now I found myself wondering if our own polar expedition would turn out to be one.

At 7 o'clock on the evening of our arrival at Nome, the helicopters began shuttling the scientists aboard from Nome's small international airport. All four helos, as the Americans called them (pronounced heel-ohs; the Canadians called them helis, pronounced hell-ees), the *Polar Sea*'s two large Dauphins and the *Louis*'s two compact Messerschmitts, buzzed back and forth over Nome's harbor like dragonflies over a marsh.

Knut and Eddy were the first to arrive on the *Louis*. Knut was the overall expedition leader. He was a natural leader, both warm and taciturn at the same time, which meant he was considerate and respectful of his colleagues and not given to temperamental outbreaks. He had been born in Brooklyn, of Norwegian parents — he is related to Roald Amundsen on his father's side — and the family moved to Norway in 1939 at the outbreak of the war. When they returned in 1947, Knut enrolled in Oberlin College, in Ohio, to study physics. After graduating, he looked around for an interesting application of physics and settled on oceanography and the Arctic. "I had been exposed to the polar literature and stories from boyhood," he said. His first trip to the Arctic was to northern Svalbard in 1956, and the experience confirmed in him a commitment to Arctic science that has remained strong ever since. "I'm in this part of the earth science business because I got started in it, and one interesting problem led to the next." He is also in it because he finds the Arctic a fascinating and challenging place for someone who enjoys field science more than lab work. "There is something to be said for making measurements, as opposed to working on the theoretical or modeling end of the spectrum." Knut finds in the physical universe a deeper contact with the sublime. "There is a remarkable satisfaction in understanding a phenomenon or a chain of events, a satisfaction that goes beyond the intellectual to something that might be described as spiritual, in the sense that I thereby feel connected to the larger Creation. In my Lutheran understanding, it is a form of grace."

Captain Phil Grandy was on the flight deck to greet the two leaders, and there was a lot of handshaking and backslapping while their gear was unloaded. Robie Macdonald and Fiona McLaughlin, colleagues of Eddy's at the Institute of Ocean Sciences, came in on the second helicopter; then Malcolm Ramsay and Sean Farley, polar bear biologists from the universities of Saskatchewan and Washington, respectively. Chris Measures, an ocean chemist from the University of

Hawaii, and Jim Elliott and Peter Jones from the Bedford Institute of Oceanography, and so on — Kathy Ellis, another ocean chemist, also from the Bedford Institute, Jim Swift from Scripps, Brenda Ekwurzel from the Lamont-Doherty Earth Observatory in New York, Caren Garrity from the Microwave Group in Ottawa — until there were 33 scientists milling around in the hangar and spilling out onto the flight deck. I imagined the scene on the American ship was much the same. It began to resemble a collage of opening sequences from M*A*S*H, except these helos and helis weren't bringing in wounded, they were landing a jubilant, even triumphant group of scientists who had been savoring this moment in the backs of their minds for the past five years. Eddy and Knut returned to the flight deck to greet the new arrivals, and there was more handshaking and hugging and laughing, until by 9 o'clock all the scientists were aboard.

Captain Grandy turned to Eddy and Knut. "If you fellas want to come up to the bridge, we're going to weigh anchor and get this show on the road."

Following the *Polar Sea* out of Nome, we rounded the Seward Peninsula and turned north into Bering Strait. We seemed to have left the birds behind, but after staring out to sea for a while I began to see them again. First one, a black-legged kittiwake (*Rissa tridactyla*) — a small gull with gray wings and clearly defined black primaries "as if dipped in ink," says Peterson — skimming a few centimetres above the water less than half a kilometre from our starboard beam, rising and falling with the waves. It was flying effortlessly and kept up with the ship as though pulled north by some magnetic force. Then my eye moved a degree ahead and I saw a second kittiwake, then a third, and suddenly I realized the ship was surrounded by squadrons of birds, all flying with the most marvelous economy, barely distinguishable from the ocean's foam-flecked surface. We were being not so much followed as escorted. I looked carefully at each one within binocular range, hoping to see a red leg — black-legged kittiwakes are common throughout the northern oceans, including the eastern seaboard, but red-legged kittiwakes (R. *brevirostris*) breed only in the

Aleutians and the Pribilof Islands. But strain as I might, I saw only black legs.

There were a few tufted puffins (*Fratercula cirrhata*) and horned puffins (*F. corniculata*). The tufted differed from the horned by having completely black underparts and two yellow tufts curving back along the sides of their heads, and those extraordinary squashed-trumpet beaks. Jim Elliott, the director of the Bedford Institute, who had demoted himself to sampling technician so that he could come along on this trip, was an accomplished amateur ornithologist, and he pointed out to me that male horned puffins also develop those yellow tufts during the mating season, and so who knew what I was looking at. I had never seen puffins of either species before, and had vaguely assumed them to belong to the North Atlantic. But here they were, at first resting calmly in the water directly ahead of the ship, then, as we bore down on them, making frantic, last-second dashes from under our bow.

We sailed due north, slipping through the narrowest part of Bering Strait, where the United States and Russia are barely 50 kilometres apart. This was the point at which Vitus Bering turned back in 1728, having proven to his own if not his czar's satisfaction that the two continents were not contiguous. Cape Prince of Wales loomed off our starboard bow, and on the Siberian or port side we could see Diomedes Island, so named in 1732 by Ivan Fedorov and Mikhail Gvozdev, the first Russians to lay eyes on what is now Alaska. They chose the name Diomedes not, I imagined, because Diomedes was the bravest Greek soldier after Achilles at the siege of Troy, but because a Diomedean exchange is one in which all the benefit goes to one side — in the *Iliad*, Diomedes trades his own brass armor for Glaucus's gold set. Now Diomedes is an island and Glaucus is a gull. The international date line, that arbitrary division between yesterday and today, passes almost directly over it. As we watched the high, round, upthrust slab of naked rock slip past our port side and sink into the ocean, it occurred to me that it was to be our last sight of land for quite some time.

Bering Strait is the narrow part of the bottleneck that decants warm Pacific water into the Chukchi Sea and eventually into the

Arctic Ocean. It is not only narrow, it is also extremely shallow, barely 36 metres deep in some places. Passing through it is almost like shooting a rapids in an icebreaker. And yet ten times more water enters the Arctic Ocean through Bering Strait than from all the rivers in Russia. Much of it is warm water. The Japanese Current, or Kuro-Si-Wo, also known as the Black Current, a Pacific version of the Gulf Stream, flows northward from Southeast Asia, past Japan, and up the eastern coast of Siberia at a rate of about 3 knots with a temperature that is 7 Celsius degrees warmer than that of the surrounding Pacific water. In hydrographic circles, 7 is a lot of degrees. By the time the Gulf Stream flows around Spitsbergen and into the Barents Sea, it is only 4.4 degrees warmer than the surrounding Atlantic water, and yet it is warm enough to keep the Barents Sea navigable nearly year-round, and to give Tromsø, on the northernmost tip of Norway, a climate similar to that of Nova Scotia. At one time, it was thought that the Japanese Current would do the same for Alaska. In the 1860s, the great, deluded German geographer August Petermann theorized that if the Japanese Current flowed through Bering Strait, it would keep the Arctic Ocean ice-free at least during the summer months, and who knew how far north? Perhaps all the way to the North Pole. This revival of the Open Polar Sea hypothesis, the centuries-old idea that the water surrounding the North Pole was somehow ice-free, was an intriguing notion and a fine example of the chilling ease with which armchair philosophy can send a great many people to their graves.

Bering Strait has always been considered the least likely route to the North Pole. Even if the Open Polar Sea were a reality, almost everyone thought that the best route to the Pole was from the east, up through Davis Strait or else north from Spitsbergen. Everyone, that is, except a French hydrographer named Gustave Lambert, who had read Petermann and who, moreover, had sailed into the Chukchi Sea aboard an American whaler in 1865 and had not seen any ice. Lambert demonstrates the danger of drawing scientific conclusions from a small sample. Upon his return to France, he began raising money for a full scientific expedition and succeeded in selling enough subscriptions in

Paris for a voyage, which he planned to make in 1869. Unfortunately, before he could sail, the Franco-Prussian War intervened, and Lambert was killed in a skirmish. The expedition died with him.

It was revived a few years later by an American naval lieutenant named George Washington De Long, who persuaded James Gordon Bennett, owner of the *New York Herald*, to buy him a ship if the navy would provide the personnel and provisions to take it to the North Pole. Only eight years earlier, in 1871, Bennett had sent one of his reporters, Henry Morton Stanley, to Africa to search for the Scots missionary David Livingstone; he was intrigued by De Long's promise of another journalistic coup. He purchased the *Jeannette* from the British navy; the ship had already seen service in the Canadian Arctic in 1875 and 1876 as HMS *Pandora*.

Captain De Long led the ill-fated *Jeannette* expedition

On July 8, 1879, the *Jeannette* sailed out of Mare Island Navy Yard, in San Francisco Bay, with a complement of 24 American naval officers and crew, after having had more than $50,000 worth of work done on her, including an ice-strengthened wooden hull and two new coal-fired engines — a sort of midlife refitting, except that, as it turned out, it wasn't the middle of the *Jeannette*'s life, but rather a few months from the end of it.

De Long justified his obsession with the Pole in a couple of scientific ways. He wasn't just going to the Pole, which would have been news; he was also going to make "observations" of the entire Arctic Ocean, which was science. He had had some hydrographic training, and another officer, the naturalist Raymond Lee Newcomb, would take care of the biology. They weren't exactly going to sail to the Pole, because De Long believed, after Petermann, that the Pole itself was located on a continent, known as Wrangel Land, a landfall off the northern coast of Siberia that had first been sighted in the 1820s by the Russian explorer Ferdinand Petrovich, Baron von Wrangel, from a point on the ice 200 nautical miles off the coast of Siberia. It was seen again by Henry Kellett in 1849 while Kellett was looking for Franklin, thinking the latter had somehow made it through the Northwest

Passage and into the East Siberian Sea (he hadn't). Wrangel Land was believed by Petermann to be connected to Greenland by a coastline running directly over the North Pole, and De Long thought all he had to do was sail into the Open Polar Sea to the shores of Wrangel Land, follow the coast north for a few hundred leagues, leave his ship anchored in a safe harbor somewhere, and continue inland to the North Pole by dogsled.

After a month and four days of heavy, sloppy steaming up the Pacific coast — loaded to the gunwales as she was with coal — the *Jeannette* arrived at St. Michael's, the former Russian settlement at the mouth of the Yukon River, in early August. There she took on more coal and two Inuit hunters, and headed north again through Bering Strait into the Chukchi Sea. At about 67 degrees N, still in open water, they turned west and worked their way along the Siberian coast, and on the last day of August they took the ice, as the whalers used to say, a few miles east of Herald Island. This made sense: the influence of the Japanese Current would decline as they sailed west, so now all they had to do was follow the ice face north to open water, and then continue across it to Wrangel Land. But after working up dead-end leads and bobbing around multi-year floes with De Long in the crow's nest and his ice pilot perched on the topsail yard, the *Jeannette* became locked in the ice on September 8, within tantalizing sight of what they thought was Wrangel Land. They had reached 71 degrees 30 minutes N. The ship had obviously settled in for a long sojourn in ice that reached higher than the wheelhouse and showed no sign of giving way to a vast polar sea of open water. Over the next few weeks, they drifted with the ice — right around Wrangel Land. This threw considerable doubt on the Petermann hypothesis: Wrangel Land was, as De Long wrote dejectedly in his log, "either one large island or an archipelago." They also reasoned that, if the massive floes of ice holding them fast were anything to go by, the influence of the Japanese Current didn't extend very far north. From a hydrographic point of view, De Long's voyage had been a tremendous success.

From the point of view of polar exploration, however, the trip was a disaster. De Long and his crew stuck with the ship for almost two years, drifting back and forth and around with the ice but never having much hope of getting out of it. On May 17, 1881, they drifted past Jeannette Island; on May 25 they drifted back again past Henrietta Island. They were drifting slowly north: on June 3, with a tired, almost grateful sigh, the *Jeannette* sank into the East Siberian Sea at 77 degrees 15 minutes N. De Long and his crew climbed into three lifeboats and headed toward shore. Before they got there, however, a fierce

gale sprang up and the boats were separated. No one knows what happened to the one captained by Lieutenant Charles W. Chipp. No one really knows how De Long's own boat reached shore. The third boat, commanded by Edward Ellsberg, the chief engineer, landed somewhere near the Lena River, and its seven men and two from De Long's boat eventually made it to a native village. The rest were never seen alive again

Ice carried fragments of the *Jeannette* across the Arctic

by anyone who knew how to inform the *New York Herald* about it.

Our first morning in the Chukchi Sea, our first real taste of the Arctic Ocean, I woke up at 4 A.M. rolling gently back and forth in my bunk as though the ship were perched on the chest of a sleeping giant. I drifted back to sleep, realizing with vague relief that I didn't seem to be sick. Charles Darwin suffered from seasickness every day he was on the *Beagle*, and his voyage lasted five years. "I am invariably sick," he wrote to a friend while in Chile. "It is no trifling evil." Science owes a lot to Darwin's condition, though, because it drove him ashore at every possible opportunity, and it was on shore that he made the observations that would change our perception of nature.

When I woke again it was 8 o'clock, and I went on deck, where I beheld a perfectly calm sea with barely a ripple disturbing the surface. The ship rolled as it moved forward, but its motion didn't seem to have anything to do with forward progress. We were riding over low hills and valleys of water, and I marveled at how closely the ocean resembled the prairies. I recalled that one way to avoid seasickness was to fix your eye on a point on the horizon, but there was no horizon, just a bank of thick fog between the water and the sky. The sound of the ship was like that of breathing, slow and deep and rhythmic, but it wasn't the sough of wind in the rigging, as I had imagined: it was the sound of air escaping from the fuel tanks through a series of hornlike safety vents welded to the deck along the outside passageway. As the ship rolled, the viscous fuel oil sloshed sluggishly back and forth, pushing foul-smelling air up through the vents. Every now and then there was a snort, and we were rocked more forcefully, as though the giant had been suddenly troubled in a dream.

I consulted my stomach, and we reached a joint decision that it was safe to go in for breakfast, with my stomach reserving the right to opt out at a moment's notice.

For meals, Stefan and I had been assigned to the officers' mess, which meant white tablecloths, sterling silver cutlery, and table service by the *Louis*'s three stewards (as opposed to trays, stainless steel, and cafeteria-style dining in the crew's mess). Meal hours were fixed according to when the crew got off their four-hour watches: 8 o'clock for breakfast, 12 for lunch, and 4 for supper. At our table, besides ourselves and three scientists, there were the radio officer, Gord Stoodley, and Steve Hemphill, one of the *Louis*'s two helicopter pilots.

Gord was a soft-spoken, friendly Nova Scotian with thinning, sandy-colored hair and a nervous mien. His mind seemed never to leave the radio room on the bridge deck, even when his body was down below stuffing itself with waffles and maple-flavored Aunt Jemima pancake syrup. Important messages could be coming in at any time, or one of the computers might have decided to self-destruct, or Captain Grandy might suddenly want a report on ice conditions farther north. This morning, said Gord, the computer connecting us with Inmarsat was acting up. Inmarsat — for International Marine Satellite — is the main communications system for ships at sea: as long as we were within range of one of its two geostationary satellites, one above the Pacific at about the equator (*Inmarsat A*) and the other over the Atlantic at the same latitude (*Inmarsat B*), then we could telephone or fax anywhere on Earth, albeit at a rate that would erase Mexico's deficit in a month and a half. We could also hook into any computer system by modem, which was how a lot of ice information came to us from the Ice Centre in Ottawa. The Ice Centre would up-link the day's ice map to *Inmarsat* A, and we would downlink it through the radio room's computer. At the moment, even though we were still within *Inmarsat* A range, our link with Ottawa was fading in and out.

"I've been trying to downlink the *Globe and Mail*," said Gord. "The Ice Centre puts it and the *Halifax Chronicle-Herald* on the computer every day, but the signal is already getting too weak. We get about half the *Chronicle-Herald* and then the link breaks off. It's as though something hasn't been loaded at the other end."

"Yeah," said Steve. "Software problem."

Holding a forkful of waffle a few centimetres from his mouth, Gord stared in glassy mystification at the forward bulkhead, where, behind sliding glass doors, rows of creamy chocolate puddings were arranged on a glass shelf. "I don't really know what's happening," he said. "We can't be getting out of Inmarsat range already."

I asked him if we were getting any ice maps for the region we'd be entering in a few days.

"Oh, no, they haven't been loading them, either," he said. "Strange."

No ice maps seemed of less concern than not getting the *Globe and Mail*, but perhaps it was too early to be worrying about ice. At $30 a try, and eight tries so far, that day's *Globe* must have been the most expensive single issue of a newspaper ever sold, and we still hadn't got it. I considered writing my first letter to the *Guinness Book of World Records*.

Steve stood up to refill his coffee cup and looked out through the starboard porthole.

"Well, well," he said. "Looks like we're getting into some ice."

3

Killer Blue

"In our Arctic language, Mr. Clawbonny, we call that an *ice-field*
— that is to say, a surface of ice which extends beyond the reach
of sight."

"Well, it certainly is a curious spectacle," said the Doctor,
"and one that acts powerfully on the imagination."

– JULES VERNE, *The English at the North Pole*

WE HURRIED OUT to the main deck to watch it come, chunks of sea ice drifting toward us like a marble armada. It came not in flat floes but in rounded scoops, wind-carved and sea-scoured into a million phantas-magorical shapes and sizes: winged porpoises, fluted obelisks, scalloped gargoyles. I had forgotten the color of ice and was startled to see it again so intensely. Where the summer sun had played with their edges, the floes were translucent crystal, a kind of candied lace that glinted oilily in the sparkling water. Deep within the larger pieces there glowed a fierce, intense emerald blueness. When a wave made one floe nod briefly and come unconcernedly up again, water streamed off its flanks in a miniature Niagara, fan-shaped and green. The snow that rode on their backs was a rich, vibrant gray, a glassy crust that glinted darkly in the sunlight. The ice bore down upon us like nightmarish chess pieces, then sailed majestically past as though we were not, after all, their target, a few at first, then a few more, until almost imperceptibly we were surrounded by a flotilla of ice, sliding

by silently without touching our ships. We felt spared. We were jubilant. We stood and watched, mesmerized; we had thought ourselves
beset by a herd of white rhinos and had seen them turn into gaz-
elles and swans. The water between chunks was eerily calm, and the
wind acquired a strange, sighing sound as it sifted through the ice's
labyrinthine channels.

Our word "Arctic" comes from the Greek *arktos*, which means
"bear." The Arctic regions were so designated by Pytheas because
they lay under the constellation known as the Great Bear. Pytheas
was a Greek mathematician who set out from Massilia in 335 B.C. to
see what lands, if any, were to be found outside those surrounding the
Mediterranean Sea. He seems to have made it somewhere north of
the Shetland Islands, perhaps all the way to Iceland but probably only
to the Orkneys. He called the region the Arctic, and the land Ultime
Thule, "the furthest point of the habitable world," and he was the first
writer in the world to describe sea ice. To Pytheas, ice represented a
kind of synthesis of all known matter: the Arctic, he said, was a place
where "neither earth, nor water, nor air exist separately, but in a sort
of concretion, like marine sponge, in which the earth, the sea and all
things are suspended, thus forming as it were a link to unite the whole
together." The land of the bear, he said, was a magical place where
the laws of nature no longer apply.

The Canadian Arctic was known to Europeans in the 11th century,
at least by hearsay, from Viking accounts of their colonizing efforts
in Greenland and Vinland — a population migration that had
been made possible by a period of global warming. The Icelandic-
Canadian scholar Tryggvi J. Oleson, in his study of the ancient history of the Canadian Arctic, *Early Voyages and Northern Approaches*
(1963), quotes this description of Greenland (albeit misidentified as
Iceland) from the 13th-century *Geographia Universalis*:

> It is called Yselandia, as the land of ice, for there the mountains
> are said to be frozen together, in the severity of the icy cold. There
> is crystal. In that region are to be found many very large and wild

polar bears, who break the ice asunder with their claws, and make large openings in it. Through these, they dive into the sea and catch fish beneath the ice.

"No less well known," Oleson continues, "were the islands of the Canadian Arctic," for it was from these, Baffin Island in particular, that the Norwegians took the white gyrfalcons that were so prized by European falconers. The Canadian archipelago was known in Arab countries, where falconry was the principal sport of the aristocracy, as the Falcon Islands. The medieval writer Abu'l-Hasan 'Ali Ibn Sa'id describes Baffin Island, which he calls Harmûsa, as lying "in the northernmost part of the inhabited world. . . . Around it lie small islands, which have falcons, and to the west of it is the white falcon island, which is almost a seven days' journey in length from west to east, and almost four days' journey in breadth. To it, and to the little northern island, men go to obtain the white falcons which are brought thence to the Sultan of Egypt."

In fact, the Canadian Arctic was so well known in pre-Columbian Europe that a globe constructed in 1492 by the German cartographer Martin Behaim depicted a nameless island in the shape and position of Greenland (with a hunter firing an arrow at a polar bear), and west of it another large island, which could only be Baffin Island, with the inscription "Here one catches white falcons." Sixteen other islands in the Canadian archipelago are shown on the globe: in the year of Columbus's departure from Spain, the Canadian archipelago was better known to Europeans than were the West Indies. Canadian falcons, narwhal tusks (sold as unicorn horns), and polar bear skins were valuable trade goods in Europe 200 years before Columbus brought back his potatoes and green peppers from the "New" World.

But no Englishmen sailed into the Canadian Arctic until 1576, when Martin Frobisher set off in search of a northwest passage to Cathay. Frobisher's chronicler, George Best, began his account of this voyage by stating what everyone already thought they knew about the North, "that during [the winter] the region must be deformed with horrible darkness and continual night. As a result, animals would not be able to seek their food, and the cold would become intolerable. Through these two evils, all living creatures would be constrained to die, and thus the region about the Pole should be desolate and uninhabited."

Nothing, said Best, was farther from the truth. The Arctic was "suitable for habitation," even in the polar zone, because in summer the sun shone for six months without setting, warming the land so much that the "infinite islands of drifting ice . . . were thawed by the great heat of summer. Only then were they driven to and fro by the tides, winds and currents to trouble the fleet. Actually," continued Best, drifting blocks of ice the size of mountains were a good sign, because "they show how great is the heat of the summer in that region when it can thaw such monstrous blocks of ice." This was the origin of the Open Polar Sea.

The North Pole itself, he added, was a temperate land inhabited by "men, women, and children and a great variety and number of beasts." Their land was "divided into four parts by four great guts or channels. These channels deliver themselves into a monstrous cavern with such violence that any ship that enters one is doomed. It cannot be held back by the force of even the greatest wind but is swept headlong into that monstrous receptacle, and then into the bowels of the earth."

After Frobisher's second voyage in 1577, Best seems to have given up trying to make the Arctic sound like the Fortunate Isles in the off-season. Perhaps Frobisher told him to lay off the tourism and get on with the adventure. The British didn't want to colonize the Arctic, they just wanted to exploit it. In the second account, the Arctic is suddenly no place for a Sunday outing. Approaching Frobisher Strait, Best writes, they "met great islands of ice, of half a mile . . . in compass," and their progress was blocked by "a continual bulwark" of ice that "defendeth the country, that those that would land there, incur great danger." Comparisons with temperate zones were this time less favorable:

> Here, in place of odoriferous and fragrant smells of sweet gums, and pleasant notes of musical birds, which other countries in more temperate zones do yield, we tasted the most boisterous Boreal blasts mixed with snow and hail, in the months of June

and July, nothing inferior to our untemperate winter: a sudden
alteration.

Ice not only prevented Frobisher from sailing around the top of
the continent to Cathay; it also made it difficult for him to land and
dig up the black ore that he had brought back from his first voyage,
and which he had convinced his backers was gold. Frobisher was selling the Arctic as a British El Dorado, 25 years before Sir Walter
Raleigh invented that myth and applied it to South America. Frobisher could land only where the ice let him:

> On July 29, about four leagues from Beare's Sound, we discovered
> a bay which was fenced in on each side by small islands lying off
> the mainland. These break the pressure of the tides and keep out
> most of the drifting ice. We anchored in the lee of a small island in
> what proved a very fit harbour for our ships. Both the island and
> the sound are now named after that Right Honourable and virtu-
> ous lady, Anne, countess of Warwick. . . . On the island where we
> were anchored was an extensive deposit of ore, in which, when it
> was washed, gold could be plainly seen.

The dichotomy is clearly set up: the only barrier between England
and Cathay, or between England and Arctic gold, was nature: and up
here, nature was ice. Ice was an indomitable foe, like the dragon of
medieval legend. Cut off its head, and it grew three more. Defeat it
one year, and it came back the next in greater force. Sail around it,
and it would close in behind you and ensnare you forever. Ice kept its
secrets hidden and fortified. Stronger than stone, sharper than steel,
more treacherous than breakers, ice created its own undertows and
back eddies. It could flip over and crush a ship in seconds, as if it were
the maw of some vast, fathomless Leviathan. Like dragons, in which
the Elizabethans still half believed, ice grew stronger as the year advanced, its crushing jaws spat fire, and to actually touch it meant
certain death, especially if you were in a wooden ship. Ice ate wood.
It swallowed ships whole. Ice gripped the British imagination and
became a living, ever-present menace, the monstrous, malevolent

guardian that stood between England and the Holy Grail: trade with the Orient.

Although dozens of British mariners sailed into the Canadian Arctic after Frobisher to search for the passage to Cathay, none found it. They did, however, find something else, something perhaps, in the end, more valuable. They discovered a new respect for nature. Captain Thomas James, for example, left England in 1631 and became stuck in the ice off Charlton Island, in the southern end of what he called (after himself) James Bay. He and his men spent a miserable winter in a makeshift cabin on the island, the first Englishmen to overwinter in the Canadian Arctic and live to tell about it. "When we stood on the shore and looked towards the ship," James wrote in an account of his voyage that became popular in England, "she looked like a piece of ice that had been carved into the form of a ship. She was packed solidly in frozen snow, while her bow and both sides were sheathed in ice. Our cables froze in the hawse-holes, and were strange indeed to behold." James's matter-of-fact prose style ("On the 19th, our gunner — who as you may remember had his leg cut off — was dying") vividly understated the horror he had lived through, and his descriptions of ice ("We had Ice not farre off about us, and some pieces, as high as our Top-masthead") inspired a passage in Coleridge's *Rime of the Ancient Mariner*:

> And now there came both mist and snow,
> And it grew wond'rous cold:
> And ice, mast-high, came floating by,
> As green as emerald.

The Northwest Passage eluded James as it had Frobisher and would so many after them. But in the course of looking for it, they got to know ice extremely well.

On the bridge, five decks above the waterline, I stood transfixed at the window, staring out at the ice through field glasses. It was still pretty loose, all first-year and brashy, but closer to the horizon I could see it tightening up into a more solid mass. We were on the periphery of the ice pack, moving easily through young ice that was being blown south into the Chukchi Sea. Ice comes in three flavors: first-year, second-year, and multi-year. First-year ice can grow to a thickness of 1 or 2 metres, but it still has a lot of salt in it and so is softer than its older siblings. An icebreaker like the *Louis* can cut through a field of first-year ice without changing gears. Second- and third-year ice gets progressively fresher, as the salt crystals leach out of it, and consequently harder as well as thicker. Sea ice, however, can grow to a thickness of only about 5 metres, because it grows downward, into the water, and 5 metres down the ocean is never below freezing. Sea ice kilometres thick exists only in the imaginations of pulp fiction writers. And those mast-high floes of Coleridge's were probably the result of two or three or even four floes piled atop one another by the pack's internal pressure, a phenomenon known as rafting. Or perhaps they were pressure ridges, two vast ice fields forced together so violently that, like continental plates, their edges were thrust upward into piles of broken blocks 20 metres high. There are no icebergs in the High Arctic; icebergs are chunks of glaciers that have calved on the coasts of Greenland and Alaska and fallen into the sea. They drift south, not north. Up here, once we penetrated the main pack, all the ice we would encounter would be multi-year sea ice, miles and miles of hard, blue ice on the move and under pressure. The early explorers called it "floating rocks"; modern seafarers refer to it as "killer blue."

Captain Grandy was standing quietly at his usual spot on the starboard side of the bridge. He, too, was absorbed in contemplation of the ice. I could see his grim face reflected in the glass of the window through which he was looking. He was a difficult man to get to know, difficult even to speak to, and not only because he was the captain. He was a man of long silences followed by great rushes of few words. When he talked, he talked like someone who has spent a lot of time alone. He had been born in a Newfoundland outport the name of which, Belleoram, is a complete mystery to lexicographers. *Newfoundland Name Lore* lists Belleoram as "very euphonious but unintelligible." But it was a decidedly seafaring village: all of Grandy's people were fishermen and mariners, and he'd grown up assuming that his life would follow theirs. "As it has," he told me, although he had to leave Newfoundland for it to do so. He went to Halifax in 1963, at the age of 16, and soon joined the Coast Guard, studying navigation at the Nova Scotia Institute of Technology during the winter between summer trips to the Arctic on the *Narwhal*, a supply vessel that worked the routes into Pond Inlet and Fox Basin with a cargo of medicine, vegetables, and mail. From there he graduated to the Coast Guard's fleet of small landing craft, which delivered their cargo to Arctic communities and weather stations by driving right up onto the beaches in an annual replay of the Normandy invasion. For some reason, most of these ships had avian names — the *Auk*, the *Skua*, the *Eider*, the *Puffin*, the *Gannet* — and Grandy worked on them until he graduated from NSIT in 1968. He then sailed as an officer aboard the *Labrador*, an old naval icebreaker that had been transferred to the Coast Guard. Since then, except for a two-year shore job in the early 1970s, he had been at sea, always on icebreakers. In 11 years as a captain, he'd been home with his family a total of 11 months.

"I love the Arctic," he said, coming out of his reverie. "The only thing that's constant up here is change."

I replied that the Arctic was then a good place to come to study climate change. He looked out at the ice, nodding indifferently.

"It's not often you get to be part of something that's taking place for the first time," he said. "When's the last time that happened? Sending someone to the moon, 25 years ago? Since then it's been one shutdown after another — shutting down the railway, shutting down the shipyards, shutting down whole towns. We're not shutting anything down up here, we're opening something up. I find that very exciting."

Our talk drifted to Canadian sovereignty in the Arctic, a subject about which he was quietly passionate. Like most Newfoundlanders, he could be disarmingly ambiguous about being Canadian, but he was in no doubt about who owned the Arctic. He'd been on the *Labrador* in 1969, when the ship was deployed to chart the waters of Viscount Melville Sound, the central corridor of the Northwest Passage, in preparation for the first-ever transit of the passage by a commercial ship — a transit made later that year by the 300-metre American supertanker *Manhattan* and a convoy of smaller vessels, including the U.S. Coast Guard icebreaker *Northwind* and the Canadian icebreaker *John A. Macdonald*. The transit raised hackles in Ottawa, where many felt it was a deliberate challenge to Canadian Arctic sovereignty. "The ship did not even fly the Canadian flag," wrote one outraged observer at the time. "Not a word appeared in the lengthy account in *Life* magazine about the role of the Canadian icebreaker the *John A. Macdonald*, which on at least five occasions had to come to the rescue of the big tanker." The *Manhattan* became stuck in ice eight times, and the *Northwind* broke a propeller and had to turn back, leaving only the *John A.* to escort the *Manhattan* to Point Barrow. The cavalier attitude of the United States toward Canadian interests and expertise in the Arctic must have been grating to a newly graduated officer aboard his first icebreaker.

American challenges to Canadian sovereignty in the Arctic continued in the 1980s, when it looked as though oil wells in the Queen Elizabeth Islands would be the answer to North America's problems with the oil supply from the Middle East. Captain Grandy's friend David Johns had been one of the "unofficial Canadian observers" aboard the *Polar Sea* in 1985, when she had sailed the Northwest Passage west from Thule, Greenland, to Point Barrow without asking for permission from Ottawa. The United States claimed (and still claims) that the passage is in international waters. Canada (the external affairs minister of the day was Joe Clark) had meekly said, No it isn't, it's ours, but, ahem, feel free to use it whenever you want. It had been another shaky moment in diplomatic relations between the two countries. Many prominent Canadians wanted Clark to assert Canada's sovereignty all the way to the North Pole. To do that, however, Canada needed a ship capable of sailing to the North Pole, and at that time Canada did not have such a ship. No one knew if it had one now. The *Louis* had been especially refitted, but there were no guarantees she would make it through the vast ice fields of the western Arctic Ocean. Even Captain Johns, when I had spoken to him in Ottawa before leaving, had given us at best "a 50-50 chance."

"It's good to be in the ice," I said. Now that we were in the pack the sea was calmer, which is why whaling captains took the ice whenever a storm brewed near the ice edge.

"It's what we came here for, I suppose."

"Would you lay odds on our reaching the Pole?"

"Oh," he said, smiling, "we'll make it to the Pole."

I asked him if getting to the Pole was a sovereignty issue.

"Well," he said carefully, "there hasn't been anything come down from Ottawa about that, but as far as I'm concerned, yes, it's there."

"Does Canada extend all the way to the Pole, then?"

"Yes," he said. "It's Canada up there."

"Will there be a race between the *Louis* and the *Polar Sea*?"

Captain Grandy chuckled. "Well, if there is," he said, "we'll take care of it."

For now, however, we were still in the Chukchi Sea between Russia and Alaska, close to where the Chukchi melded into the Arctic Ocean proper. Our plan was to work our way north in a kind of zigzag pattern, one ship following in the wake of the other, stopping every few degrees to carry out our various science programs, intending to reach the North Pole sometime toward the end of August. After that we would zigzag back south again, along a different route and doing more science, until we reached Point Barrow on the 21st of September. That way we would cover as much of the unknown part of the Arctic Ocean as possible, and get back in time to refuel, if we had to, before the winter ice pack joined the shore ice and cut off our escape route to the Pacific.

The science was to be carried out in a series of stops, or science stations, each to take place over a specific feature on the floor of the Arctic Ocean. To the Coast Guard crews, we were sailing along a flat, ice-clogged sea that would do everything in its power to prevent us from going where we wanted to go. But to the scientists, we were traversing a terrain, one that existed, at the outset, barely 200 metres below our hulls but that gradually, as we moved out over the edge of the continental shelf, deepened to a depth of more than 4 kilometres, and consisted of long, towering mountain ranges called ridges, huge, flat plains known as basins, and deep, chasmlike holes, or abysses. The really big abysses were called abyssal plains. Each of these submerged features not only contained deep-ocean deposits that recorded their own sedimentary history going back millions of years in a few centimetres; they also affected the current patterns and composition of the water that flowed through and above them, exactly as mountain ranges and prairies affect the air that flows above the earth. At each of nearly 100 science stations, teams of scientists and technicians on both ships would drop instruments into the ocean at precisely calculated depths in order to sample everything from the rock below the benthic mud, and the mud itself, to the water at specific levels between the mud and the ice. Other scientists would study the ice, and still others would sample the air above the ice and even the amount and quality of the sunlight as it traveled through the air, through the ice, and into the water.

Each of these samples would be brought aboard and either sealed in containers and stored in the holds for later study, or, in the case of the more volatile substances, tested immediately in the sophisticated laboratories that had been set up throughout the labyrinthine confines of the ships. Each team of researchers had had containers of equipment loaded onto the two ships and had spent the three days since leaving Nome frantically working to get the equipment unpacked and the labs set up in time to begin the sampling work when we got far enough north.

I spent the time gathering information. I wanted to get a sense of what global warming was before everyone started looking for signs of it. And I had 70 experts at my disposal with nothing to do but tinker with equipment and talk to me about their theories and experiments. After a few days, I felt I was beginning to grasp the enormity of the problem and the range of experimentation that would be going on for the next two months of the voyage.

Global warming is not by definition a bad thing. The planet has survived numerous climate changes over the course of its 5-billion-year history: in fact, multicellular life (i.e., ours) was made possible by global warming. Four billion years ago, when the sun burned 25 to 30 percent less vigorously than it does today (it is progressively flaring itself out like a struck match, but don't throw out your sunscreen yet), the earth was too cold to support life. Life requires water in all three of its phases — liquid, vapor, and solid. Four billion years ago, all the water on Earth was solid. But then intense volcanic activity is thought to have filled the atmosphere with gases — especially methane, ammonia, and carbon dioxide — that acted like a blanket and allowed the earth to warm up to habitable temperatures.

Global warming and cooling events have coincided with massive shifts in human history and may help explain the development of the various cultures, or even so-called races, that inhabit the earth. Early hominids first migrated out of Africa into Europe and Asia during a period of global warming. The first humans to cross into North America were able to do so because global cooling had turned Bering Strait into an ice road, and subsequent periods of glaciation pushed them down into South America. The Vikings, as I've mentioned, were able to settle in Greenland at a time when what is now a singular mass of rock and ice lived up to its name — although the average global temperature was only half a Celsius degree warmer than today's, the sagas speak of green pastures and large flocks of sheep. The Inuit already there were able to move farther north to avoid (or perhaps to plan) clashes with the Norse invaders. While there, they hunted whale and walrus populations that had also migrated that far north for the first time. In the Middle Ages, during what is known as the Little Ice Age, global cooling drove Germanic peoples south and led to the overthrow of the Roman Empire in 375 A.D. The role of climate in human affairs has been studied for some time; the role of climate *change* is only being appreciated now that a new phase of global warming has appeared on our horizon.

The greenhouse effect — the analogy by which we explain why the surface of Earth is warmer than, say, the surface of Mars — is the engine that drives global warming, and therefore is not by definition a bad thing either. Solar radiation from the sun strikes the earth's surface; some of it is absorbed into the earth, the rest of it is reflected back toward space in the form of thermal radiation. When the earth had no atmosphere, all the reflected radiation bounced back into space. When the earth gradually became surrounded by quazillions of molecules of various energy-absorbing gases, they collectively prevented a large proportion of that thermal radiation from leaving the atmosphere. (The greenhouse analogy was first proposed in 1827 by French physicist Jean-Baptiste-Joseph Fourier, who likened Earth's atmosphere to the panes of glass in a greenhouse, which allow short-wavelength solar radiation into the greenhouse but prevent long-wavelength thermal radiation from leaving it; result: hot air and azaleas.) Each of these molecules acts like a miniature Earth — the rays of long-wavelength thermal radiation hit the molecule, some of their heat energy is absorbed by the molecule and the rest of it is reflected off at a mathematically computable angle. Some of this reflected radiation is aimed back toward Earth and makes the surface a little warmer than the sun's primary radiation would by itself: about 33 Celsius degrees warmer. The average surface temperature of Earth now is about 15 degrees C; without the greenhouse effect, the average would be about –18 degrees C, too cold to support what we refer to as life-as-we-know-it.

What we are experiencing now, what we call global warming, ought to be called an enhancement of the greenhouse effect. There are too many molecules of the wrong kind of gas in our atmosphere. The atmosphere is composed of many different gases, but the one with the greatest influence on the greenhouse effect (after water vapor) is carbon dioxide, or CO_2, one atom of carbon electrochemically fused to two atoms of oxygen to form a single molecule. CO_2 is better at absorbing long-wavelength thermal radiation than any other gas; it makes up only about 0.04 percent of the atmosphere, but it is responsible for about 55 percent of the the greenhouse effect, which means that you don't have to tamper with the CO_2 level very much to make a huge difference in the amount of heat thrown back at the earth.

In what scientists have begun to refer to as an "unperturbed world" — an imaginary Earth not threatened by an enhanced greenhouse effect — the amount of radiation entering the atmosphere is in a state of equilibrium with the amount leaving it. Until recently, the amount of carbon dioxide in the atmosphere remained more or less constant; with minor fluctuations, it had remained at around 280 parts per million (ppm) for the past 11,500 years or so. We know this from bubble samples taken from ice cores drilled into the glaciers in Greenland

and Antarctica: in a sample from a segment of ice that was carbon-dated at 130,000 years old — a period of interglacial global warming between two ice ages — the carbon dioxide level was 400 ppm. The lowest CO_2 concentration revealed by the ice core was dated 40,000 years ago, when the carbon dioxide level was only 150 ppm; that was during the Wisconsin Ice Age, when the average global temperature was about 7 degrees c and the ice sheet was thickest and had advanced to its southernmost point — the northern tip of Long Island. Lower the amount of CO_2 below about 200 ppm, and you get an ice age; raise it above 400 and palm trees start growing on Baffin Island. Today's CO_2 level is 365 ppm and rising.

Oxygen is a by-product of life, not the other way around — there was no oxygen in the earth's atmosphere until there was life, in the form of blue-green algae, to exhale it — and carbon dioxide is the by-product of death. When a tree burns, it gives off carbon dioxide. When vegetation rots, it gives off carbon dioxide. CO_2 can also be produced nonorganically — by the erosion of limestone, for example. Normal fluctuations in atmospheric carbon dioxide can be accounted for by such processes: an inordinate number of forest fires erupt in Siberia, and up goes the carbon dioxide level; a plague of insects wipes out a significant portion of the Amazonian rainforest, huge sections of it begin to decay, and again, up goes the CO_2 index. If these phenomena occur at the same time, enough carbon dioxide might enter the atmosphere to trigger an episode of global warming.

"Trigger" is the word here because large-scale global warming is not a single event but a series of linked events, each one triggering a sequence of new ones: higher temperatures cause greater water evaporation, which creates more cloud cover, which traps more infrared radiation, and so on. In scientific jargon, this is called positive feedback. Negative feedback discourages the global warming chain reaction: higher temperatures might permit faster forest regeneration, for example, which would use up some of the excess atmospheric carbon dioxide, which would lessen the greenhouse effect and lead to global cooling.

By and large, global warming and cooling events have taken place on massive scales. The series of mass extinctions that marked the boundaries between the three dinosaurian eras — Jurassic, Triassic, and Cretaceous — all ended with global climate changes that wiped out up to 90 percent of life on Earth. And they occurred at more or less regular intervals. Someone has tried to work out exactly how regular those intervals are, with a view to proving that our current global warming is merely a natural episode on a colossal, knowable timetable, but without much convincing effect. No matter how you look at it, the enhanced greenhouse effect we're experiencing now is not normal. We are causing it, by burning fossil fuel and by cutting down or burning inestimable tracts of the world's forests. We increase the carbon dioxide load by decreasing the number of plants that absorb carbon dioxide during photosynthesis, and by releasing carbon dioxide into the atmosphere.

Think of a barrel of oil or a lump of coal as liquid or solid carbon dioxide that has been trapped inside the earth for millions of years. When plant material decayed to make coal during the Carboniferous period, for example, its carbon dioxide was not released into the atmosphere but was sealed up underground, which is why the plant material petrified into coal instead of composting into soil. When we dig down into the bowels of the earth, extract coal or oil from the vaults in which they have been sequestered, bring them to the surface, and set fire to them, we are finally releasing into the atmosphere carbon dioxide that was produced 240 million years ago, adding it to the carbon dioxide that would naturally be released during our own time. That is what we have been doing, at an ever-increasing rate, since James Watt perfected the steam engine in 1769 and Sir Richard Arkwright applied it to textile manufactories in 1790. Or at least since 1859, when Jean Joseph Étienne Lenoir invented the first workable internal combustion engine and mounted it on the back of a horseless carriage.

After 11,500 years of equilibrium, the atmospheric carbon dioxide level began to rise around the middle of the last century. Before the Industrial Revolution, atmospheric CO_2 was about 280 ppm; in 1958, the first year anyone thought to measure it in real time, it was 315 ppm. By 1989, when the early warning system for global warming started going off, it had jumped to 360 ppm and was rising at the rate of

1.8 percent per year. Although the actual amount of CO_2 entering the atmosphere dropped slightly from 1989 to 1990 (because of the recession, not because of any industrial or government reaction to scientists' concerns about global warming), it began to rise again in 1991. It is rising still.

At its current pace, the carbon dioxide count should reach double the pre–Industrial Revolution level — that is, 560 ppm — by the year 2070. If that happens (and it will, if not by 2070 then almost certainly by the end of the next century), the earth's average temperature will have risen by about 5 Celsius degrees. A 5-degree average global temperature rise means almost no increase at the equator but an approximately 20-degree rise at the North Pole. That means, at best, a seasonal ice cover, much as we now get on the Great Lakes. We will be living, as Eddy put it, "on a different planet," a planet less like Mars and more like Venus, whose average surface temperature is 450 degrees C and whose atmosphere, not coincidentally, is made up almost entirely of carbon dioxide.

4

The Lower Depths

The warmly cool, clear, ringing, perfumed, overflowing, redundant days were crystal goblets of Persian sherbet; heaped up — flaked up, with rose-water snow.

— HERMAN MELVILLE, *Moby-Dick*

O N OUR SECOND DAY in the ice, Stefan and I were flown by helicopter to the *Polar Sea*, where we were to spend the next five days. We were greeted on the flight deck of the American ship by Lieutenant (Junior Grade) Karen Arnold, who took us below to show us where to stow our personal gear and Stefan's incredibly bulky video recording equipment, and then gave us a brief tour. "We've been outfitted for special science work," she said as she led us through the *Polar Sea's* dimly lit passageways below decks. I was amazed at the difference between the *Louis* and the *Polar Sea*, which is essentially the difference between the two Coast Guards. The Canadian Coast Guard is part of the civil service, a branch of the department of transportation; the American Coast Guard is an arm of the military — the fifth service, it's called. Canadian Coast Guard ships are unarmed and look like floating office buildings: wide corridors, lots of light and air, linoleum decks and wood paneling, nicely appointed dining rooms and curtained lounges. U.S. Coast Guard ships look like battleships at war: their passageways below decks are barely wide enough for two people to pass, and after 6 P.M. they are kept dark, with only the occasional red light glowing from the bulkhead, presumably so that no lighted porthole would give away the ship's position to the enemy. Although all U.S. Coast Guard ships are called cutters — a quaint historicism that means a one-masted sailboat rigged like a sloop — the *Polar Sea*'s official designation is WAGB-11, radio language for Coast Guard Gun Boat 11, and I was assured that if trouble broke out and the *Polar Sea* were required to go into battle, large deck-guns in the hold could be brought up and mounted in minutes.

The officers' mess was a small, sparse room with two long tables that accommodated 20 officers each, Mike Powers, the executive officer (shortened to XO), at the head of one and the captain, when he showed up, at the head of the other. Each table was covered by plastic tablecloths, and the food was passed around, not served by a bevy of stewards. Scientists were welcome to eat anywhere they wanted, and most preferred to eat below in the crew's mess, which was more informal and closer to the labs. She was a very republican ship. The plates were of unbreakable Melmac. There was no wood paneling in the officers' mess or anywhere else. The chairs and tables and even the doors were made of steel. As one of the officers explained, "Wood splinters when it's hit by torpedoes. During World War II, more sailors were killed by flying wood than by any other means, including drowning."

After dinner, Karen told us that Captain Lawson W. Brigham had invited us up to his quarters for an official welcome. His quarters were one deck up from the mess and one deck down from the bridge. His door had a sign on it that read: "Knock and Enter: Please Wipe Your Feet." We knocked and entered a small entranceway and wiped our feet on a bright red carpet. Directly ahead was the captain's private galley, and in it the captain's private steward, sitting on a blue milk crate reading a back issue of *Military Wife*. When he saw us come in he jumped to his feet and hurried out to greet us. His blue Coast Guard uniform was starched and his name tag said Grant.

"How y'all gettin' on?" he asked. "Cap'n Brigham's in through heyah."

He showed us into the captain's dayroom, a large, bright, comfortable office with a sofa and two easy chairs, a wooden coffee table under the glass top of which was a schematic diagram of the *Polar Sea*, a long mahogany boardroom table with eight padded wooden chairs, a massive oak desk, and, running the length of the forward bulkhead below a row of white-curtained portholes, a set of bookshelves made of solid wood. Captain Brigham's sleep was clearly untroubled by dreams of torpedoes and flying splinters. The walls were covered with maps, paintings of the *Polar Sea* and other Coast Guard cutters, letters of congratulation from various political figures (all in wooden frames), and a large, framed photograph of two polar bears growling playfully into each other's open maws.

I ran my eyes along the books. Most dealt with the Arctic: Arctic exploration, Arctic navigation, and especially shipping in the Northern Sea Route, the Russian version of the Northwest Passage, a tantalizing link between Murmansk and Asia that Vitus Bering had been searching for in the 18th century, and that Russia had been trying to keep open for commercial shipping since 1938, when the Soviet icebreaker *Sibiriakov* made the first — and very nearly the last — transit of it in a single season.

There was Krypton's *The Northern Sea Route and the Economy of the Soviet North*; a beautiful edition of Burney's *Chronological History of North-Eastern Voyages of Discovery*; and Starokadomskiy's *Charting the Russian Northern Sea Route: The Arctic Ocean Hydrographic Expedition, 1910–1915*. There was also a collection of essays, entitled *The Soviet Maritime Arctic*, published in 1991 by the Woods Hole Oceanographic Institution, edited by Captain Brigham himself. It was an altogether spectacular collection. I could feel my fingers beginning to itch.

Captain Brigham must have noticed. "Feel free to borrow any of these that interest you," he said. "Come in any time. As you can see, I've made something of a study of Arctic icebreaking, especially as it pertains to the Northeast Passage. I've been up there many times. What a fascinating subject it is. Think of it: for 60 years the Soviet Arctic was the most tightly controlled and inaccessible region in the world. Now, since the breakup of the Soviet Union, the Northern Sea

Route could become a new international waterway linking Europe and the Far East, cutting 20 days off the shipping time from Oslo to Tokyo through the old Suez Canal route — not to mention linking Siberia's incredible natural resources with the rest of the world. If only they could keep it open more than three or four months a year. The Russians now have 107 icebreakers in service, 20 of them Polar-class and nuclear-powered, and 300 ice-strengthened freighters. I've seen them. I've been on them, plowing through ice in the Kara Sea in convoys of up to 12 ships — six icebreakers and six ice-strengthened tankers. And still the pressure of the ice stops them in the shallow water and narrow passages between islands. Nowhere to push it, you see."

He was a neat, eager, energetic man in his early 60s, friendly, attentive, and shy. When he smiled, which was often, he closed his eyes, and the light coming in through the portholes glinted off his glasses. His dayroom had the distinct air of the professor's study, a subtle but pleasant effect created by the smell of books and wax, dust motes floating in shafts of filtered sunlight, and the assumption that knowledge for its own sake mattered a great deal. I was struck by the difference between the two captains. On the *Louis*, I rarely saw Captain Grandy except on the bridge; never in the labs, rarely in the common rooms. He had not invited me anywhere for an official welcome, seemed barely aware of my presence. Captain Brigham, as I would learn, was rarely on the bridge. He left the day-to-day running, the actual navigation, of the ship to his officers and spent his time in his dayroom or on the lower decks, chatting with the scientists.

"I'm delighted to be taking part in this scientific expedition," he said as we sat down with coffee and cookies. He leaned back in his chair and rested his head against a needlepointed cushion depicting a black-and-white terrier. "I've had some training as an oceanographer myself, even done some teaching at the Coast Guard Academy, and I think I have a good rapport with the scientific community on board. I understand their needs and indeed can anticipate many of them. I think the role of the captain in today's Coast Guard is changing rapidly — I've just spent four years in Washington on a special committee on the future of the Coast Guard in general. We're no longer required to be the fiercely independent, Captain Ahab types who have to manhandle our ships through the ice by sheer willpower alone. With all this modern technology at our disposal, getting a ship from point A to point B is pretty much a matter of pressing the right buttons. I see the role of captain as having evolved into something more akin to that of mayor, someone who facilitates working relationships among the various communities aboard his or her vessel, between the scientists and the crew, for example, or the crew and the officers."

I remarked that the two captains on the voyage seemed to complement each other, that one was experienced in Arctic ice and the other was more attuned to the expedition's scientific goals.

"Yes," Captain Brigham nodded. "Phil and I are great buddies. I have a lot of respect for his abilities as a mariner. But he and I are clearly from different worlds. Getting to the North Pole, for instance, is more important to him than it is to me. We've already been there, for one thing," he said, meaning the United States, "at least by submarine." The USS *Nautilus* had crossed the North Pole under the ice in the 1950s; like Amundsen's passing over the Pole in an airship, the submarine's admittedly impressive feat is not looked upon as an actual conquest of the Pole. "Getting there would be nice for me personally, of course," continued Captain Brigham, "because it would mean that the *Polar Sea* will have made it to the world's most southerly navigable spot — 79 degrees 29 minutes and 95 seconds south — in the Bay of Whales off the Ross Ice Shelf, where we were just before our deployment up here to the Arctic, and to the world's most northerly navigable position, the North Pole, in the same year. That would be something. No one has ever done that before." He took a contemplative sip of coffee. "But getting to the Pole?" he said, shrugging. "Once you get to 85 degrees, the rest is a piece of cake."

"Can I quote you on that?" I asked lightly, taking out my notebook: 85 degrees was the farthest north ever reached by an American surface vessel: in 1991, the *Polar Sea*'s sister ship, the *Polar Star*, had had to turn back from an attempt to reach the Pole by the eastern route. She had broken one of her propellers in the ice.

"Well, okay," he said, laughing, "maybe 88 degrees. Anyway, once you're in the vicinity, you might as well pop over and see what it's like at the Pole. But for me, getting to the Pole is secondary: it's the scientific aims of the expedition that are paramount."

Fortunately, I thought, the two goals weren't mutually exclusive. But I couldn't help wondering what would happen if they came into conflict, if the two captains had to choose between achieving the expedition's scientific goals and reaching the North Pole.

The expedition had two captains; it also had three science leaders. Knut Aagaard was the expedition leader. Eddy Carmack, on the *Louis*, was the Canadian science leader. For them, science was of course the main reason for the trip, and there was important science work to be done at the Pole. Oceanographers had compiled "sections" — profiles of the various layers of water within the ocean from the surface down to the bottom — of every part of the world ocean except the western Arctic. Making a transit to the Pole would literally bring those sections full circle.

The American science leader aboard the *Polar Sea* was a geologist. Art Grantz was with the U.S. Geological Survey, and his specialty on this trip was ocean-floor tectonics. His headquarters was in Menlo Park, California, but he looked as though he spent more time in the field than chained to a desk. His features were as tough and craggy as the Alaskan rocks he had spent most of his life studying; he was also one of the most admired and respected scientists on the expedition. He'd been roaming the interior ranges of Alaska since 1949, coming to his own recondite opinions about their formation and subsequent history. In 1990, he had edited Volume L of the Geological Society of America's monumental *Geology of North America*; his volume, called *The Arctic Ocean Region*, included two of his own articles: "The Canada Basin" and "Geology of the Arctic Continental Margin of Alaska," classic studies. As far as Arctic geology went, Art was at the top of the escarpment. What he saw from there was chaos.

The problem was, Arctic geology didn't go very far. Once you get out off the continental shelf and out of the Beaufort Sea, away from areas of intense oil exploration, not much is known about the formation of the ocean floor or the climatic history delineated in its sediments.

That hasn't prevented a lot of people from voicing a lot of opinions about it. How and when the Arctic basins and ranges were formed were matters of intense and sometimes vituperative speculation.

Over the years, three theories for the formation of the Arctic sea floor have received widespread support. The earliest, dating from the 1930s — long before the general acceptance of the theory that the continents ride around on top of floating tectonic plates — held that the area now covered by the Arctic Ocean was once a mountain, like the area now covered by Antarctica. This might be termed the parallelism hypothesis: if one end of the earth is marked by a bulge, so must the other end be. Over countless millennia, this Arctic mountain eroded onto the Canadian and Eurasian continents, wearing down gradually until it disappeared altogether, in fact forming a declivity into which water from the surrounding continents then drained to form the new ocean. This was the classical view of the geologic process, in which continents were mountains midway through the process of becoming oceans, and it was defended into the early 1970s, when knowledge of plate tectonics rendered it untenable: continents simply did not behave like that.

According to the theory of plate tectonics, continents are huge slabs of rock floating on vast internal seas of molten lava. They drift around on this liquid core in much the same way huge floes of ice drift around on the Arctic Ocean, much more slowly but still subject to the whims of wind and currents, occasionally and catastrophically colliding with one another. When that happens, they push up huge piles of crumpled debris at their edges, which at the rims of continental plates we call mountain ranges. If the pushing continues, the edge of one plate will ride up on top of another plate, just as two floes will raft together when they collide, and material that is on top of the subducted plate will pile up at the point of contact, as if the upper continent were a huge bulldozer blade and the new material simply accumulated in front of it. On the west coast of North America, we call this new material, between the Rocky Mountains and the Pacific Ocean, various things — California, Washington, Vancouver Island.

For some reason, in the past several million years most of these continental plates decided to get together in the Northern Hemisphere, crowding shoulder to shoulder like football players in a huddle, with their feet dangling down into the Southern. The crowding created a triangular open area at the top which, according to the new theory, became the Arctic Ocean, beginning approximately 175 million years ago and ending 50 million years ago — quite recently in the geological scheme of things. What is now Bering Strait was once a much wider gap, so wide that there was virtually no distinction between the Pacific and Arctic Oceans. About 50 million years ago, a continental plate to the south, called the Kula plate, pushed its way into the gap and turned the Arctic Ocean into a closed sea, its only entrance and egress through the Greenland Sea; eventually, the Kula plate became subsumed under the larger continental plates, the gap that is now Bering Strait reopened slightly, and the Arctic Sea became the Arctic Ocean.

This seemed a satisfying explanation of things until a third hypothesis, the one in favor now, came along. This was the hypothesis Art favored. In the 1950s, William S. Carey, a farsighted geologist from Hobart, Tasmania, who anticipated plate tectonics before just about anyone else except Alfred Wegener (who came up with it at the turn of the century), looked at a map of North America and couldn't help noticing how the Rocky Mountains and the Coast Mountains veered suddenly westward when they reached Alaska, as if they had once been straight but something had happened to bend them toward Siberia. He suggested that some kind of sea-floor spreading must have taken place in the Svino Chasm, north of the Brooks Range but still in Alaska, pushing the mountain ranges off line.

Carey's hypothesis met with some resistance, probably because he had come up with it without ever having been to Alaska; he'd simply sat in his armchair in Tasmania and looked at a map. But a handful of geologists — among them Art Grantz, who knew Alaska the way few people know their backyards — agreed with Carey enough to take his proposal seriously. In 1966, the pivotal year in the acceptance of the plate tectonics paradigm, Grantz provided solid stratigraphic evidence in support of Carey's hypothesis. Some kind of fan-shaped sea-floor spreading between the Eurasian and North American continental plates, he said, had pushed Alaska's mountain chains into the

Pacific Ocean, opening up the Canada Basin in the process. The pivot seemed to have been located near the mouth of the Mackenzie River.

"If you look at a map of the Arctic Ocean," Art explained, spreading out just such a map in the *Polar Sea*'s officers' mess, "you'll see that it's shaped like a triangle, which is why the entrapment theory at first looked so attractive. But there's a better explanation for it." He put a salt shaker on one corner of the map and a plastic bottle of Heinz ketchup on another. The ketchup bottle fell off the table and onto the deck. "We're a bit cramped for space here," Art apologized. "This isn't a five-star hotel like the *Louis*."

Art's story of how the Arctic Ocean was formed begins in the North Atlantic. There, the Atlantic Mid-Ocean Ridge, a rupture in the ocean floor, runs up the east coast of Greenland, through Fram Strait, and across the top of the Arctic Ocean, where it changes its name to the Arctic Mid-Ocean Ridge. This ridge is scar tissue, built up over the gap made by sea-floor spreading between North America and Europe about 55 million years ago. As the American and European plates drifted apart, the gap between them bubbled up with molten lava from the earth's core. This lava hardened to form a ridge. When the plates drifted apart again, more lava bubbled up, until eventually the ridge became much higher than the surrounding sea floor.

The Atlantic Mid-Ocean Ridge pushed North America and Europe apart; farther north it moved Greenland away from Spitsbergen, forming Fram Strait. As it moved even farther north into the Arctic region, it sliced off a thin section of the north coast of what is now Siberia and pushed it out into the Arctic Ocean, where it is now known as the Lomonosov Ridge, an 80-metre-wide mountain range that rises 2,500 metres from the ocean floor and runs almost directly over the North Pole. If you could walk a kilometre below the ocean's surface, you could walk along the Lomonosov Ridge from the northern tip of Greenland to the New Siberian Islands in the Laptev Sea, crossing over the Pole, following almost exactly the shoreline of the mythical continent of Wrangel Land, as imagined by August Petermann. During the middle Cretaceous Period, the shore of Wrangel

Land was the northern edge of Asia: Petermann had been right all along, but his timing was off by about 125 million years.

This movement of the Lomonosov Ridge away from Siberia divided the Arctic Ocean into two massive basins, known as the Eurasian and Canadian Basins. Within the Canadian Basin, there is a smaller deep plain called the Canada Basin. This, says Art, was created by a spreading of the North American continent, a kind of unfolding of the Arctic shoreline that separated what is now Alaska from the Northwest Territories. Imagine a closed pair of scissors with the Mackenzie Delta as the pivot. Now open the scissors, with the coast of Canada moving to the east and the coast of Alaska moving to the west, pushing the Rockies in Alaska, in fact pushing all of Alaska, out toward Siberia and creating a gap between the blades into which rushed the waters of the Canadian Basin.

"That much is known and agreed upon," Art continued, "although it's still hypothetical." In fact, the Canadian Basin is the largest piece of the earth's oceanic crust for which there is no rigorous plate tectonics explanation. That is largely because for most of the year the basin is covered with 2 metres of ice, and in the 1960s and '70s, geologists had to rely on drifting ice stations and the occasional science-oriented submarine for their deep sediment and crust samples. "We had some data from those sources," said Art, "but it was pretty spotty. We couldn't really get a complete picture." Spotty is right: until the 1980s, only nine core samples had been taken from the Arctic Ocean floor, three by scientists camped on ice stations and six from the *Nautilus*.

In the past few years, however, Art and a team of geologists from the U.S. Geological Survey had taken ocean-bottom readings of the area between the Northwind Ridge in the west and the Canadian islands in the east.

"We pulled up all kinds of neat stuff," Art said. "Fossiliferous limestone and dolostone from the late Paleozoic, Triassic marine shales and siltstones, Jurassic and Cretaceous bedrock, middle Cretaceous lutite. You name it, we found it. The neat thing was that everything we got on the Northwind Ridge was identical to strata in the Canadian archipelago, suggesting that at one time the ridge was a part of the North American continental shelf." Finally, after 25 years of theorizing, Art had solid evidence to support Carey's hypothesis.

And in the process he had come up with an entirely new hypothesis of his own.

Our cabin was more like a bunkhouse at a summer camp than a stateroom on the *Lusitania*. There were 150 crew members aboard the *Polar Sea*, more than twice as many as on the *Louis*. Space was a problem. Below decks, quarters were divided into four-person, six-person, and eight-person cabins. Ours was a six, three tiers of metal bunk beds bolted to the metal deck, with metal curtain rods around each set, a metal-encased reading lamp above the pillow, and metal lockers with metal doors that didn't quite shut but made a lot of noise trying. If we were struck by a torpedo, I thought, we wouldn't be killed by flying splinters, but the noise would turn us into gibbering idiots. The room was always either in total darkness or lit by an eerie red bulb that made me feel part of some top-secret submarine experiment. I undressed in the dark, climbed up to the top bunk on a metal ladder, drew the curtain, turned on the lamp, and tried to read.

"Antarctic sea ice," I read in an article by Barry Lopez in *Orion* magazine, "differs from ice in the Arctic — the pressure ridges are fewer and less formidable; the snow cover is deeper; and the ice is not as thick, not as hard." Lopez had sailed with the American icebreaker *Nathaniel B. Palmer* on her maiden voyage to the Weddell Sea in 1993. Antarctic sea ice is always first-year ice; it forms and melts, forms and melts, growing faster and thicker than first-year Arctic ice, but never lives long enough to become the really thick, hard, multi-year ice that we would soon be dealing with up here. "It's less obdurate, less wrinkled," continues Lopez, "but it is as anchoritic — it imposes a stillness and a terrestrial vastness over the water."

The noise from the *Polar Sea*'s engines throbbed through the ship, through the metal bunk, through the felt pillow, and into my head; I'm an embryo, I thought, whose mother has heart palpitations, gastric gurglings, serious intestinal disruptions, and lousy posture. A high-pitched chirping sounded at regular five-second intervals through the ventilator shaft, as though some robotic bird had been welded to the main deck under one of the funnels. Something hard and round, a forgotten cannonball perhaps, rumbled back and forth along the deck directly above my head with every pitch and roll of the ship, thundering into tin canisters full of ball bearings and careening into galvanized steam fittings, which then hissed like furious snakes.

In the morning, I found myself standing in front of one of six shower stalls in the men's washroom, awaiting my turn. Behind me, men barely old enough to shave were shaving. Before me, affixed to the metal door of the stall, was a photocopied notice giving instructions for the economical use of the facilities:

SHOWER ON: 15 SECOND RINSE
SHOWER OFF: SOAP
SHOWER ON: 15 SECOND RINSE
SHOWER OVER.

Apparently, wartime rationing was in effect.

"I just hope to God this thing works," said Kent Berger-North. He was sitting on the metal deck of an unheated loading bay, hunched over a rat's nest of red and black wires spraying from a small black box. He chose two wires and soldered them together. He had been working on the black box in exactly that position for the past two hours.

"Science is fun," he muttered quietly to himself as I passed. It was 2 o'clock in the morning, still broad daylight, and we'd been stopped in the Chukchi Sea, at 67 degrees 47 minutes N, since midnight, waiting to complete the first science station of the expedition. Nothing was going right. "Science is fun," he repeated, like a mantra. "Science is fun."

The black box housed the trip mechanism for the biology team's carousel water sampler, also known as the rosette, which was standing in disarray in the center of the loading bay. It was supposed to be dangling at the end of a cable, 10 metres from the bottom of the ocean, electronically opening and closing a ring of plastic bottles collecting samples of sea water. But when Kent had assembled the rosette to test it before letting the winch haul it over the side, the bottles hadn't tripped. He had taken off each in turn and checked its stopper, eventually tracing the problem to the switch assembly in the waterproof black box. He didn't quite know what the problem was — "It just wouldn't talk to us," he said — but figured if he rewired the whole thing the problem would go away.

"Anything I can do to help?" I asked.

"You can put the bottles back on the rosette," he said.

The bottles, called Niskin bottles, looked like extra-large gray plastic coffee Thermoses with rounded lids at each end that could be opened or closed by electronic signals from the ship. The signals traveled through a wire in the core of the winch cable, into the black box, and thence to the bottles. I snapped each bottle vertically into clamps around the rosette's metal frame, hoping that I had got them all right side up, and when I had finished, the assembly looked like a playground carousel about 2 metres in diameter, with a cable coming out the top and 24 evenly spaced cylinders arranged around its periphery. Attached to the bottom of the frame were three sensors about the size of small electric motors. One was a CTD, an instrument that measures the conductivity, temperature, and depth of the water, the most basic measurements in oceanography; conductivity is determined by the water's salt content and so is actually a measure of salinity. The second instrument was a fluorometer, which shines a fluorescent light into the water that illuminates the various microscopic creatures that inhabit it, phytoplankton and such; the third sensor was a transmissometer, which detects the phytoplankton illumined by the fluorometer and, 25 times a second, sends a signal up to the surface that tells the scientists at the control switch whether the rosette is traveling through just plain water, or water peppered with phytoplankton.

Stefan was outside the loading bay, talking to Delphine Thibault, a marine biologist with the Bedford Institute who had come to study zooplankton in Arctic waters. Zooplankton is a general term for an array of tiny marine animals; some feed on herbaceous phytoplankton and others are carnivores, dining on even smaller animals such as flagellates and ciliates. Delphine wanted to know what kinds of zooplankton lived in the Arctic, what they ate, and how many of them there were in an average cubic metre of surface water. To do this she had a large net, about 3 metres long and a metre across at the wide end narrowing down to a small, weighted container at the thin end, which collected whatever got in the net's way as it was hauled up. A winch was supposed to lower the net over the side to a depth of 50 metres — the deepest extent of light-dependent life in the ocean — and then haul it aboard with its microscopic catch. But the winch wasn't working.

"It's good to get all the bugs out of the system now," said Delphine.

Lisa Clough was having better luck on the fantail, the stern end of the ship one level below the flight deck. Lisa was a benthic biologist from East Carolina University; her province was the bottom of the sea, where life is anything but light-dependent. Her instrument is known as a box core, a large metal guillotinelike affair with a square box at the bottom and a cantilevered scoop that swings shut when the box hits the bottom, trapping sediment in the box. The frame is then hauled back onto the ship, the box hopefully full of about a cubic metre of sediment going down 40 centimetres into the ooze.

"It doesn't always work," said Art Grantz, looking down into the water where the box core's cable disappeared into the murk. "Sometimes the scoop doesn't swing all the way under the box, and you lose the sediment on the way up. Sometimes it hits a rock. Sometimes the whole frame goes down sideways and the box doesn't even get into the sediment. You can never tell what's going to happen down there. When it comes up, you look into the box. If you see mud and a whole bunch of little worms looking back at you, you've got a successful box core. If all you see is a lot of soupy sea water, you do it again, if you have time. We're in only about 50 metres of water here, so we'd probably have time. But later on, when we're in 3,000 or 4,000 metres and the box core takes a couple of hours to go down and a couple more to come up, you only get one shot at it."

The fantail winch was working, and when the box core was raised out of the water and lowered onto the deck, we hurried forward and peered anxiously into the box: it was satisfactorily filled with mud, a fine, grayish-black ooze mixed with small pebbles, tiny shells, and little squirming sea creatures — burrowing anemones, copepods, sea snails, and worms — topped with a few centimetres of sea water. When the box was removed from the frame and dragged away from the edge of the deck, Lisa knelt down and went to work on it. She took four glass tubes and pushed them through the water into the sediment. Then she fixed a glass lid with a rubber seal to the top of each tube and reached down into the mud to slide lids under the tubes' bottom openings, warming her hands in her armpits between tubes. The temperature of the water was 1.7 degrees C.

"They're not quite airtight yet," she said. "When I get them in the wet lab, I'll have to remove every air bubble in there with a syringe. It takes hours. But air bubbles would supply surface oxygen to the samples, and since the idea is to measure how much oxygen these critters consume under normal benthic circumstances, we can't have any extraneous oxygen getting in and messing up our numbers."

When she got all the oxygen bubbles out of the water, she would measure the amount of oxygen in it, leave the cylinders in a freezer for 24 hours, and then measure how much oxygen was left. This would tell her how much had been consumed by the critters in the mud. Then she would open the cylinders and pass the mud through a fine-mesh net, identify and count the critters, and get an idea of how much oxygen is required by each critter.

"We're doing some pretty basic natural history work," she said, "just finding out what animals are here and at what rate they're taking up oxygen. Hopefully, we'll be able to come back sometime and do follow-up studies, to see if that rate is changing, and if so, how that change correlates to other changes we observe in the climate."

The box core was a modern refinement of an instrument designed by John Ross, one of the first polar explorers, who in 1817 devised something he called a Deep Sea Clamm. Before Ross, scientists assumed there was no life at the bottom of the ocean — not enough air, no light. Ross proved them wrong: he brought back sealed jars of benthic mud from Davis Strait that contained shells and worms very similar to those Lisa was pulling up now and gave them to Sir Joseph Banks, the most eminent biologist of the day. But Ross was not popular as an explorer; he had turned back at the ice face in Frobisher Bay, and his report was virtually unread. It was left to some later genius to reinvent the Deep Sea Clamm and to call it a box core, and to use it to continue Ross's work after a hiatus of 150 years. The Arctic is like that.

Lisa pulled the cylinders out of the box, placed them upright on a plywood tray, and carried them into the wet lab, passing through the loading bay where Kent was still soldering away at his switch assembly. It was 5 A.M., and both the rosette cast and the plankton haul had been called off. Station 2, at 70 degrees N, was scheduled for 10:30 the next morning, and Kent had to get the rosette ready and try to get some sleep before then.

He looked up as we stepped over his legs, dripping sea water and mud onto his jeans. "Science is fun," he said.

5

Fram's Wake

We were nearly surrounded by ice, which closed in the ship on
all sides, scarcely leaving her the sea-room in which she floated.

    – MARY SHELLEY, *Frankenstein*

E MERGING from the biology lab onto the main deck
was like coming out of a cinema in the middle of the
afternoon. I looked at my watch: 1 A.M. The mostly
open sea around us, sprinkled here and there with small chunks of
ice, was a brilliant dark blue, the sky pale and wispy, the sun not five
points above the horizon. A steady breeze blew from the north. Before coming out I'd checked our position on the GPS monitor: 69 degrees 40 minutes N. In a few hours we'd be stopping for station 2. I
took out my binoculars. The northern horizon was rimmed with a
thin white line hovering just above the water: ice blink, the bright reflection of ice against the sky. Whalers in these waters would look for
ice blink to tell them where not to go, unless a gale was brewing. I
scanned the horizon for a break in the line of white reflection, which
would indicate open water, a lead in the solid ice. I didn't find one.

Art Grantz was in the geology lab, a narrow, gunmetal room filled
with electronic equipment, bank upon bank of toggle switches, wires,
computer screens, and digital readouts, their red numbers racing at
binary speed and apparently at random, since the ship was stopped.

The stacks reached from deck to ceiling and left only nominal space between them for us seemingly irrelevant human beings. Art, seated, was flanked by two ocean-bottom profiling machines called line scan recorders.

Art and his team had been in these same waters almost exactly a year earlier, aboard the *Polar Star*, and had made some intriguing discoveries. "Geological studies in Alaska and Canada," they'd written, "indicate that the Canadian Basin was formed by rotational rifting of Arctic Alaska away from Arctic Canada about a pole of rotation in the Mackenzie Delta, beginning about 125 million years ago. Little geological or geophysical data, however, exist from the basin itself by which this hypothesis can be tested."

Art was hoping to procure that information during this trip, but he was also planning to accomplish much more. Within the Canadian Basin, sprinkled about the otherwise featureless plain, are a number of features that ought not to be there if the basin had been formed by simple rotational rifting: the spreading of the scissors would not have left mountainous peaks, for instance, between their widening blades. And yet there they were, the Northwind Ridge to the west, along the edge of the Chukchi Borderland; and, at the other end, closer to the Pole, the Alpha and Mendeleyev Ridge systems. How did they get there?

Art thought they may have been moved into the basin along a transform fault line after the actual rotational rifting had taken place. "I think," he said, as the electronic pencils of the line scanners clicked away behind his head, "and I emphasize that this is my theory and I can't prove it, but then no one has been able to disprove it, either — I think the ridges in the Canadian Basin are extensions of the Arctic Mid-Ocean Ridge. Come out in the hallway and I'll show you what I think happened."

Taped to both walls in the narrow passageway were lengths of computer printouts with dark, squiggly lines running along them — the track line of the 1993 *Polar Star* cruise — and a large bathymetric chart of the Arctic Ocean, showing the submarine basins, ridges, abyssal plateaus, and plains that make up the bottom of the Arctic Ocean. Art ran his finger along a line between Greenland and Iceland, over an area marked Denmark Strait.

"This," he said, "is the Atlantic Mid-Ocean Ridge. Ridges don't always move in straight lines. You see how this one was moving nicely northward here, and then suddenly" — his finger stopped north of Iceland, at the Jan Mayen Fracture Zone — "it takes a hop to the right for a few hundred kilometres, then continues north again. When a ridge takes a hop like that, we say it hits a transform fault, a crack in the earth's crust that runs at right angles to the direction of the ridge.

"Well, I think the same thing happened again when the mid-ocean ridge hit the Asian continental shelf up here in the Laptev Sea: I think it took a sudden hop over to the middle of the Canadian Basin, pushing the Northwind Ridge and the Alpha-Mendeleyev systems out into the basin. Then it continued up into Alaska. As I said, that's what I think happened, but I can't prove it. I haven't been able to find that transform fault."

Art thought he knew where to look for it, though: behind the Northwind Ridge, at the western end of the Chukchi Borderland, where on the map a series of jumbled-up bathymetrics looked suspiciously like the right-angled ridges and troughs of the Jan Mayen Fracture Zone. Art would love to get the *Polar Sea* positioned over those features so he could train his seismic sensors on the ocean floor below and find out what was down there. In fact, he had persuaded Knut and Eddy to alter the expedition's original track line to get him and his equipment over the borderland. If the rock on the east slope of the Northwind Ridge turned out to be the same as the rock on the Siberian continental shelf, then the likelihood is that a transform fault had moved it there. In 1993, a year of "exceedingly favorable ice conditions," according to Art's report, the *Polar Star* had taken him over the zone along two track lines, one at 78 degrees and another at 79 degrees, and he thought his seismic sensors had picked up a hint of a high-angle fault system deep beneath the ocean floor that might have been the fracture zone he was looking for. This time, he intended to take a longer, closer look.

"I call it Charlie," he said, running his finger on the map over the jumble of underwater contour lines. Penciled over the area was Art's proposed track line for this trip, with a series of numbered dots representing science stations dedicated to the geology program. "The

Charlie Transform Fault," said Art. "That's where I'll find it. Right there."

In the loading bay, the rosette was assembled and ready to go, but there was no one about. Delphine's plankton net was folded neatly outside the sliding doors. I found Kent in the crew's mess, drinking coffee and staring at an old Abbott and Costello movie on the TV monitor mounted high in one corner. The *Polar Sea* had 1,500 video movies on board, and someone seemed determined to go through all of them. Kent had finally got the rosette ready, had gone to bed, and had been up since midnight preparing for the upcoming science station.

"So when is it going in?"

"It's not," he said.

"Is the switch still not working?"

"The switch is working fine," he said. "The winch is not working."

"Oh," I said.

Kent sighed. "This trip is cursed," he said.

There are those who have felt that the whole western approach to the Arctic Ocean is cursed. Certainly George Washington De Long must have thought so as he sat in the *Jeannette* for two years watching her being ground to sawdust by the ice. And the American nature writer and environmentalist John Muir agreed with him. Muir sailed into these waters in July 1880 aboard the U.S. Coast Guard ship *Corwin*, which had been sent to search for the *Jeannette* and two missing whaling ships.

Affixed to the *Corwin*'s bow was a pointed steel plate, like the blade of a snowplow, which was called an "icebreaker." The device had been used before in the south, to open channels in iced-in harbors and to push ice away from bridge abutments, but no one had thought to bring one along on an Arctic voyage before. The icebreaker worked marvelously; if the *Jeannette* had had one, she might have been found. But, as Muir notes in his account of the *Corwin*'s search, she was not.

The *Corwin* cut through first-year ice all the way to Wrangel Island (at the time still called Wrangel Land), where Captain Hooper landed a shore party and raised the American flag. The party built a cairn and placed in it copies of the *New York Herald*. But although the men tramped around Wrangel Island for several days (proving it was an island, not a continent) before heading back to Alaska, they found no trace of either the *Jeannette* or her crew. "A land more severely solitary," wrote Muir when he had returned to San Francisco, "could hardly be found anywhere on the face of the earth."

Five years later, however, a group of Inuit hunters near Julianehaab, on the west coast of Greenland, found some curious items frozen into slabs of sea ice that had drifted ashore, and turned them over to the Danish colonial manager, a man of scientific bent who published an account of the find in the *Danish Geographical Journal*. It was a fruitful bit of beachcombing. The items included a list of ship's provisions signed by Captain G. W. De Long, a manifest of the *Jeannette's* lifeboats, a pair of oilskin trousers, and the peak of a U.S. naval officer's cap with the name F. C. Lindemann stamped on the brim. In ancient times, there was believed to be a lake near the top of Mount Strello, in Portugal, where the wrecks of ships lost at sea floated to the surface; something like that had materialized now in Greenland, with objects from the *Jeannette* that had been lost in the East Siberian Sea, 5,000 kilometres away, having somehow crossed the Arctic Ocean to turn up on the west coast of Greenland.

One of those who read about the find was Henrik Mohn, a Norwegian professor of geography who speculated, in the Oslo newspaper *Morgenblad*, that the items must have "drifted with the ice over the Polar Sea," exiting the Arctic Ocean through the Greenland Sea and circling around under the southern tip of Greenland. If so, Mohn reasoned, then their trajectory must have taken them very close to the North Pole. This article was in turn read by the young Norwegian explorer Fridtjof Nansen. Nansen had already sailed into the Arctic aboard the sealer *Viking*, and at the time of Mohn's article he was preparing to make the first crossing of the Greenland ice cap on skis — a feat he and five companions, including Otto Sverdrup, accomplished in 1888. Perhaps because he had read the *Jeannette* article, Nansen collected samples of "dirty ice" — drift ice with sediment trapped in its frozen matrix — when he arrived on Greenland's western shore, and determined that the sediment in it had come from the coast of Siberia. He realized that if dirt and pieces of a ship could drift across the Arctic Ocean caught in the ice, perhaps an entire ship could, and perhaps she would drift over the North Pole.

Nansen may also have got the idea from his acquaintance with whalers. In 1883, the artist William Bradford had published an account of a meeting between De Long and a half-dozen whaling captains in San Francisco, just before the *Jeannette* sailed out of San Francisco and into oblivion. De Long wanted the whalers' assessment of his chances of finding the Open Polar Sea and getting to the North Pole. As Bradford later recalled,

Fridtjof Nansen designed the *Fram* to drift over the Pole

the whalers "all gave their opinions, mainly upon the point of greatest interest, the probable direction of the winds and currents at the time when Lieutenant De Long expected to reach Wrangel Island. But there was one among them who kept ominously silent, not venturing an opinion or offering a suggestion."

The quiet man was Captain William Nye, one of the most experienced New England whalers in the American fleet. After everyone had delivered themselves of their lengthy and often contradictory advice to De Long, Bradford turned to Nye: "Captain Nye has not given his opinion, and we would like to hear from him."

"Gentlemen," Nye replied, "there isn't much to be said about this matter. You, Lieutenant De Long, have a very strong vessel, have you not? Magnificently equipped for the service, with unexceptionable crew and aids? And you will take plenty of provisions, and all the coal you can carry?"

To each of these questions De Long replied affirmatively.

"Then," said Captain Nye, "put her into the ice and let her drift, and you may get through or you may go to the devil, and the chances are about equal."

If Nansen had read this account, he may have taken his idea from William Nye as early as 1883, two years before the *Jeannette* material turned up in Greenland. If so, the next paragraph must have given him pause: "Poor Captain Nye!" wrote Bradford. "He ventured in there after Lieutenant De Long, into those same Arctic regions, in the prosecution of his enterprise as a whaler, and was never heard of again."

If the Greenlanders discovered the flotsam from the *Jeannette* on the day it arrived on their coast, it had crossed the Arctic in 800 days; Nansen calculated that it would take two and a half years for a ship deliberately made fast in ice to work her way over the Pole and out through the Greenland Sea to open water. But he also knew that no ship then built could withstand the pressure that the pack ice would impose on its hull. The *Jeannette*'s hull had been strengthened to a thickness of 48 centimetres with Oregon pine and American elm, reinforced with cast-iron bands, and braced athwart with 30-by-30-centimetre beams; according to its fitters, it was the most "ice-resistant" hull ever built, practically unsinkable. Still, she had been ground to kindling by the relentless Arctic ice in a little more than 18 months.

So Nansen set about revolutionizing Arctic ship design. He and the Scottish marine architect Colin Archer came up with a ship, the *Fram* (Danish for "forward"), whose rounded hull, rather than being crushed between two floes, would be pinched up on top of the ice like a pumpkin seed between thumb and forefinger. In fact, whaling vessels had been built along these lines for years — the whalers said that such a hull had been "built flaring," and it is entirely possible that the *Viking* had been just such a ship — but Archer took the flaring concept to new extremes. The *Fram* was shaped like a 402-ton melon, with a flat bottom so that she wouldn't roll over when she lifted up onto the ice. Nansen and his crew steamed out of Oslo in the summer of 1893 and immediately discovered why ships with rounded hulls had become unpopular some 900 years earlier. They rolled too much. The *Fram*, top-heavy and melon-hulled, pitched and bobbed like a channel marker. Never before had a crew so looked forward to getting caught in the ice. They rounded the top of Norway and rolled west across Siberia until they reached Novosibirskiye Ostrova (the New

Siberian Islands), just south of where the *Jeannette* had disappeared. Then, on September 25, at about 79 degrees N, Nansen gratefully took the ice, nudging the *Fram*'s blunt bow between the floes and shutting down her engine. He raised a wind generator (another of Archer's innovative ideas) and sat back to wait for the ice to carry them gently over the North Pole.

The *Fram*'s rounded hull prevented her from being crushed by ice during her two-year drift

A year and a half later, he was still waiting. In January 1895, he made some fresh calculations: they were drifting back and forth a lot, but their northward progress was only about 1 degree of latitude, or 60 nautical miles, every five months. At that rate, he figured, it would take a little over eight years for the *Fram* to reach open water again on the other side of the Pole. As if that wasn't discouraging enough, they weren't even drifting in the direction of the Pole; they were slanting off on a course that would take them no higher than about 83 degrees, more than 400 nautical miles off their goal.

In early March, Nansen left the ship and most of her crew, hoping to walk to the Pole. He took along one companion — his Danish lieutenant Frederik Hjalmar Johansen — 28 dogs, enough food for three months (if you include the dogs), and two kayaks. The crew remained with the ship making hydrographic observations that, in many instances, are still the only ones we have. Nansen and Johansen, meanwhile, didn't get very far north — by April 8 they had struggled over 20-metre pressure ridges and kilometre-wide leads to get to their highest point, 86 degrees 13 minutes N. Rather than turn around and go back to the *Fram*, which after a month of erratic drifting would have been almost impossible to find, they struck directly for Franz Josef Land, hoping to run into Wrangel Land along the way (De Long and, later, the *Corwin* had already proven that Wrangel Land didn't exist, but Nansen still believed there was an archipelago of islands at the North Pole). After nearly three months of contending with broken ice, widening leads, sub-zero temperatures and starvation, they reached Franz Josef Land at the end of August, in time to build a hut, as Nansen described it, out of stones scraped into a wall surmounted by a walrus skin stretched over a piece of driftwood. The two men survived the long Arctic winter, keeping themselves alive on gull meat and two lamps that burned a small amount of walrus blubber. The following spring, 1896, they emerged from the hut like grubs from under a log and paddled feebly over to a smaller island, where they ran into a British polar expedition led by Frederick Jackson. Jackson's party was sponsored by Lord Northcliffe, the British newspaper baron, and the meeting between the Victorian gentleman, dressed as though he had just popped out for a spot of grouse shooting, and the bedraggled, half-starved Nansen, clutching his putrefying bit of gull, was the kind of photo opportunity the world had not seen since the *New York Herald* had arranged a similar encounter between Stanley and Livingstone. Unfortunately, there were no photographers around to take advantage of it.

The *Polar Sea* and the *Louis* had been shrouded in fog all afternoon, but it lifted after supper and Stefan and I were shuttled back to the *Louis* while the vessels were stopped for station 2. The scientists on the Canadian ship were also busy; the *Louis* had a rosette and a box core of its own, as well as several deep-sea Challenger pumps for contaminant sampling, and two plankton nets, affectionately called "bongos" by the biology team.

Now that we were in ice, lowering equipment over the side was a delicate operation. We could drop the rosette through an open space between two floes but, because we and the ice were drifting, we couldn't guarantee that the hole would still be there when it was time to haul it back up. Leads were continually opening and closing around us. There was some danger that one of these small floes would snap the winch cable, and raising the rosette or the box core into the bottom of even a small floe could damage its sensors.

Captain Grandy had stopped in a narrow lane of open water between two gigantic fields of ice, so that the rosette and box core could be safely lowered. The black water stretched ahead of us like a high-way cutting through flat, snow-covered fields. The kittiwakes, which had followed us all the way from our last sight of land, had been invisible during the fog but now circled above the water, seven of them, heads bent way down, like white vultures scanning for roadkills. Every so often, one would raise its wings above its body and plummet head first into the water, disappearing into a splash of white and emerging with a small arctic cod in its beak. Then, like as not, a black pomarine jaeger (*Stercorarius pomarinus*), the raptor of the Arctic Ocean, would dive at the kittiwake and try to wrest the fish from the smaller bird's beak. Groups of ivory gulls (*Pagophila eburnea*) sat on the ice like marble statues, as white and still as doves, watching the aerial combat and rooting, or so I imagined, for the under-gull.

Beside me at the rail was Ikaksak Amagoalik, one of the two Inuit guides Eddy had invited on the expedition for much the same reason he had invited me; he believed that good science is a synthesis of the widest possible range of points of view. Ikaksak's black eyes cast an amused light on everything he saw. A short, wiry hunter in his mid-40s, he had been born in Pond Inlet, on the west coast of Hudson Bay, in 1948, and when he was eight years old, his and two other families had been "resettled" to factory-built houses in Resolute, a small community on Cornwallis Island. The resettlement was part of an earlier government's attempt to assert sovereignty over the Arctic. According to the Geneva Convention, no country can claim sovereignty over territory occupied by a nomadic people, and so the Inuit had been compelled to move into such permanent townsites as Resolute and Grise Fjord. Eight other families were moved to Resolute from northern Quebec, said Ikaksak. "We had different languages completely," he said. "We couldn't understand each other, even though we were both Eskimo. But we taught them some of our language, and we picked up some of their language, and now our language is all mixed up."

"Did you know where you were going, then?"

"No, no idea. But I guess Mom and Dad knew what was going on."

"What did they tell you?"

"It's been 40 years," Ikaksak said. "I don't think I can remember everything."

We watched the ice for a while. Behind us, Mary Williams, an ice scientist from Memorial University in Newfoundland, had laid an ice core out on a workbench and was drilling a series of holes into it. Into each hole she inserted a probe; then she wrote down the internal temperature of the core as it varied from one end to the other through the thickness of the floe from which it had been taken.

"Give me your binoculars, quick," said Ikaksak.

I took them off my neck and handed them to him. He put them to his eyes and stared at a dark object on the ice, about half a kilometre away.

"What do you see?"

"I think it's a seal," he said. "Nope. Just a hunk of ice." He handed me back my glasses and laughed. "My eyes ain't so good any more."

"Eddy tells me you've been to the North Pole already."

"Yes," he said. "Four times."

"Four times!"

"Once by plane, once walking, and twice on snowmobiles." He leaned on the rail and turned toward me. "The last time was a few years ago — a group of Norwegians hired me to guide them to the Pole. I'm pretty good with machines, and I'd been there before. Maybe they thought I was lucky, I dunno."

"Was it a hard journey?"

"Not too bad," he said. "When we got close to the Pole, maybe around 87 degrees, there were lots of places where the ice had piled up on top of itself. Some of those ridges were 15 or 20 feet high. We had to cut through them. We didn't have a chain saw with us, just my ice chisel, which I made myself and which I always carry. It's a one-inch steel bar about 7 feet long, with a flattened end, very sharp. It took me sometimes six hours to cut through one ridge, and then we would go a short way and there would be another ridge. We had to get everything down nice and smooth, right to the bottom, so those snowmobiles could go over with the sleds. It was very hard. I had to be careful not to cut myself with the chisel. One day we traveled for 12 hours and made only 9 kilometres. Sometimes I would say to myself, This is not the place I want to be."

Ikaksak looked out at the ice.

"But you made it to the Pole," I prompted.

He smiled and shook his head: "They said we did. But me, I think we missed it by about half a kilometre. I don't know. The GPS satellite doesn't work so well when you're that close to the Pole. The Norwegians wanted to go back and forth a few times to make sure, but I said, No, this is close enough. I didn't want to go over to the Russian side.

"So we started back, heading south. Well, up there it was south everywhere, but we had to go back through Norway. It was already May, and the ice was breaking up. We were going very slowly, because when we got to a floe we had to wait for another floe to come up close so we could get the snowmobiles across. After a few days I could see we weren't going to make it, the floes were getting smaller and too far apart, so we called on the radio for a plane to come and get us from Spitsbergen. Then I had to look around for a floe big enough for the plane to land on. I knew it needed to be at least half a metre thick and about 500 metres long. Finally I found one, and we got the snowmobiles over to it, but when the plane landed we had to leave a lot of stuff behind. Two snowmobiles, the komatiks, lots of food and cans of gasoline. We left it all there, on the ice. All we took was one snowmobile, to put in a museum!"

He laughed like a man who enjoyed laughing.

Mary Williams had been a member of a six-person team that had attempted to ski to the North Pole in 1986. Her group had left Spitsbergen in late January, when the ice pack was reasonably solid and the temperature was supposed to have stayed in the −50s. But 1986 was one of those climate-change years, a winter of exceptionally high temperatures everywhere, including the Arctic Ocean, and the ice had begun to break up early. They made it to 86 degrees, about 550 kilometres north of Spitsbergen, before turning back.

The first person to make it to the North Pole without flown-in support (if you discount the dubious claims made by Frederick Cook and Robert Peary in 1909) was the British explorer Wally Herbert, who got there by dogsled in 1968. There had been many other attempts, but no one before Herbert (and very few since) had done it without outside help. The year Mary made the attempt, another group had set out from the northern tip of Ellesmere Island, a Norwegian team using snowmobiles (not the group led by Ikaksak). "I've talked to pilots who said they'd flown them over ridges and leads," said Mary, "picked them up in one place and set them down a few miles farther on. I don't think that counts. Wally had supplies dropped to him, but he made it every inch of the way on his own." I thought how incredible that the first human beings got to the North Pole just one year before the first human landed on the moon.

I watched Mary finish drilling holes into her ice core and recording its temperatures. She then cut it into thin slices, which she called "pucks," with a circular saw, bits of ice flying everywhere like gray sawdust. She dropped each puck into a Ziploc bag, and I helped her carry the bags down to the ship's walk-in freezer. Later, she would measure their grain size, chart their crystal structure, then thaw them and measure their salinity. Every measurement would contribute to her knowledge of how a warming climate was affecting ice thickness and strength. She had identified five different zones within the core: a top layer of well-aged snow overlying a thicker zone of snowy ice, ice that had melted slightly during the Arctic summer and subsumed some of the snow on top of it. Beneath the snowy ice was a thin layer of "bubbly polycrystalline ice," in which the ice was mixed with air bubbles and crystal structure was churned up and irregular. Below that, the main section of the core was solid, columnar ice — hard, white ice whose crystals were regularly arranged, whose grains were of uniform size, with almost no brine remaining between them. This was the solid core of the floe, ice that was supposed to remain unmelted for years until it drifted into the North Atlantic. Beneath this was another layer of softer, slightly warmer ice, where the floe had been in contact with the sea. Here the ice was riddled with brine channels, drainage holes through which the floe had leached most of its salt crystals back into the water. This was the layer that would change the most if the temperature of the water beneath it was altered.

Mary wanted to know how hard this ice was to break. "I guess our ultimate goal," she said as we stacked the pucks in the freezer, "is to be able to make a kind of safety map of the Canadian Arctic, something that will indicate to navigators which ships can go into which zones and when." To do this, she had to be able to identify the various structural properties of ice. In the freezer, which was set at –20 degrees C — far colder than the air outside — she had set up an instrument that would measure the flexural strength of a chunk of ice. The ice was laid out on the instrument's table, downward pressure was applied to a point along its surface, and a meter measured the precise amount of force required to snap the ice in two. This force would be different for ice of different ages, salt content, temperature, grain size, and density.

"We don't yet understand why there is a correlation between those four properties," said Mary. "We know that as one changes, they all change. Temperature and salinity gives brine volume, for example, and if the temperature changes — as it does as you go down the column — then so does the amount of brine. And brine volume correlates with the ice's strength. But we don't really know why. At least we're at the point where, if we know the temperature and salinity of a particular ice floe, we can calculate how much force a ship will need to break it." What she needed, then, was to find some way to predict the temperature and salinity of specific ice types, say a 2-metre-thick floe of multi-year ice, so that she could assign a force requirement to it. That would tell a ship's captain how much horsepower his ship would need to get through a particular hunk of ice.

"With a ship like this and a map like that," said Mary, closing the heavy freezer door as we left, "getting to the North Pole would be a piece of cake."

6

Of Ships
and Ice

I doubt if ever a boat was blessed with a more beautiful and at the
same time a more capable bow. . . . To touch that bow is to rest
one's hand on the cosmic nose of things.
— JACK LONDON, *The Cruise of the Snark*

A HORIZON of ice seems to exist outside time. It is space
without dimension, a glare of white relieved only by
the geometry of floe and ridge. In fact, though, there
is history everywhere in the Arctic. The ice acts upon the mind like a
blank wall, inviting cartographers to scribble names across it like graf-
fiti. Sailing up Bering Strait and still within sight of land, we had
been surrounded by the ghosts of explorers and scientists who had
preceded us: Vitus Bering, Georg Wilhelm Steller (Bering's naturalist), Baron von Wrangel, Sir John Barrow, and the literary gamut
from the *Odyssey* to the *New York Herald*.

But even now, far from land, there were echoes of the Arctic's
human history. As I stood on the *Louis*'s deck I counted seven kittiwakes, diving like arctic terns into our wake. Four pomarine jaegers
circled above them, sometimes diving down to meet an unsuspecting
kittiwake as it came out of its dive. I watched one encounter with my
field glasses: the jaeger missed the fish and took off after the kittiwake,

then plucked one of the kittiwake's tail feathers with its beak, as if in frustration, and spat it out while continuing the chase. I watched an ivory gull drift silently by, without looking down at us. It passed so close to the ship I could almost count its wing feathers and see its stern, eaglelike expression, like an archangel delivering an important message. Then Jim Elliott pointed to a lone gull flying well behind the ship.

"Look," he said excitedly, "a Sabine's gull."

I had been hoping to see a Sabine. They were named after Edward Sabine, the physicist and astronomer who, in 1818, accompanied the first expedition to set out with the stated intention of finding the North Pole — the *Dorothea*, commanded by Captain David Buchan, and the *Trent*, under a young naval lieutenant named John Franklin. Taking the Greenland Sea route, they made it only as far as the northeast tip of Spitsbergen (80 degrees 40 minutes N), where the *Dorothea* had to enter the ice to avoid the high waves from a fierce gale. When the storm subsided, the ship had sustained such damage that she and the *Trent* had to turn back. As a voyage of discovery it had been disappointing, but as a scientific venture it gave us Sabine's gull — albeit indirectly, since the bird was actually named later by the scientist's brother, Joseph, in honor of Edward's Arctic ordeal. The voyage also gave us Franklin's gull, named after John Franklin, which, like Sabine's and Bonaparte's, has a black hangman's hood.

The *Louis's* fantail is a lonely place, conducive to such gloomy thoughts. Two decks below the main, almost at water level, it is a small, open, railed-in platform under a low ceiling, the sole purpose of which, it seems, is to provide a handy jumping-off spot for anyone contemplating the abyss. The most striking view is of the churning wake, stretching out behind the ship's three gigantic propellers like rails behind a moving train; an image of infinity. Looking down into it is like peering into a maelstrom, or a raging cataract. And looking up, above the roiling water, is worse: endless ice and limitless sky. Between us and the North Pole were 1,200 nautical miles, or 1,392 statute miles, or 2,240 kilometres of solid, multi-year, pressure-ridged, rafted, hummocky ice to plow through, and then plow through again to get back. It occurred to me to wonder how much ice our two icebreakers were capable of breaking.

In 1989, when Eddy had first called Captain Johns about the proposed expedition, the Canadian Coast Guard did not have a ship capable of breaking through that much ice. None of its six largest active icebreakers at the time was big enough for a polar attempt. They were named after figures from Canadian history: the *John A. Macdonald*, the *Pierre Radisson*, the *Sir John Franklin*, the *Des Grosseillers*, the *Henry Larsen*, and the *Norman McLeod Rogers*. (Norman McLeod Rogers was a supply minister in Prime Minister Mackenzie King's wartime cabinet who died in an airplane crash in 1942. The airport in Kingston, Ontario, where I live, is also named after him: every time I fly out of it I wonder how an airport came to be named after a politician who died in a plane crash.) All except the *Larsen* were old and badly in need of repair. The *John A.*, the pride of the Canadian icebreaker fleet for 30 years, the model used by the Soviets when they built their first nuclear icebreaker, was sold for $1 million in 1993 for scrap metal. The *Rogers*, built in 1969, was sold to the Chilean navy: renamed the *Almirante Oscar Viel*, she is a pawn in that country's gambit for a section of Antarctica. The *Sir John Franklin*, built in 1979, spent most of her life in Arctic service, escorting tankers and other less capable ships through the Canadian archipelago, but for the past few years had been employed almost exclusively in the Gulf of St. Lawrence. I had been aboard her briefly in 1989 during a series of ice-margin experiments; she was a well found, cozy ship accustomed to accommodating ice scientists and their often peculiar needs, such as sailing into rather than around ice, but she was definitely tired.

An icebreaker takes a lot of abuse; its life expectancy is about 25 years, and it generally requires a midlife refitting after a dozen years in service. The *Franklin*'s original diesel engines were so inefficient (they burned up to 150 tonnes of fuel per day in heavy ice) that she would never have been able to carry enough oil to get to the North Pole and back.

When Eddy spoke to Captain Johns, he had asked for the *Henry Larsen*, and over the next four summers Johns arranged for the IOS team to make several trial runs on the ship into the Canada Basin, partly to do oceanographic and hydrographic work but also to develop methods and skills for the expedition. They soon realized that the *Larsen* was not the ship for a polar crossing. For one thing, she had only a Class 3 rating. The rule of thumb is that a Class 3 ship is capable of cruising steadily through 3 feet of ice: anything thicker and the ship has to rev up to icebreaking mode, which nearly doubles her fuel consumption. A Class 3 icebreaker is admirably suited for breaking through first-year ice in the lower Arctic or the Gulf of St. Lawrence but is too light for the kind of expedition Eddy was planning: she has only two propellers, and her diesel-electric engines can generate only about 16,000 horsepower. If one propeller were broken by heavy ice, the remaining one would never get the ship back to open water. The *Martha L. Black*, a smaller icebreaker but also rated Class 3, became stuck in ice north of Point Barrow in September 1988, broke one of her two propellers trying to free herself, and had to be rescued by the *Polar Star*. It had been an embarrassing incident, and one the Canadian Coast Guard would rather not repeat in a mission as important — and as public — as an attempt on the North Pole.

There were two other possibilities in Captain Johns's mind, but neither was very realistic in 1989. One was the mythical super-icebreaker the Canadian government was still promising to build, a dedicated scientific research vessel so large and powerful that she would be able to take a team of 50 scientists anywhere in the Arctic, at any time of year. In fact, she was being designed to spend four years in the Arctic without ever needing to touch land. She was going to be the largest icebreaker in the world: larger than the United States' largest icebreakers (the *Polar Sea* and the *Polar Star*); larger than the Soviet Union's nuclear-powered icebreakers. She was to be called *Polar 8* — the 8 indicating her ice classification; a ship that could cruise through 8 feet of ice without being aware of it.

The *Polar 8* was announced by the Conservative government in 1987. She was to be built in Vancouver at a cost of $320 million, a juicy contract for a province that had elected 19 Conservatives in 1984 but looked as though it would lean toward the New Democrats in 1988. For the next three years, the Coast Guard and the science community hummed with excitement over their new supership; when I sailed through the Northwest Passage on the *Martha L. Black* in 1988, the *Polar 8* was all anyone talked about. Canada was going to have the biggest and best Arctic research vessel in the world. She was going to have two captains — Pat Twomey and Phil Grandy — alternating every four months. She was going to remain in the High Arctic for four years at a stretch, with crew changes and fresh scientists flown in every month by helicopter. She was going to have the latest in hull design and navigational equipment. She was going to have a gymnasium. She would establish Canada's claim to sovereignty in the Arctic, from the continental shelf to the North Pole, once and for all.

She was going to be nuclear in 1987, and so would need refueling only once every three years; she was going to have five gas turbines in 1988; then she was going to have diesel engines, but really efficient ones. In 1987, she was going to have two long-range helicopters with top-of-the-line on-board microwave radar imagers for ice reconnaissance — no reliance on weak communications links with the Ice Centre in Ottawa; in 1988 she was going to have two medium-range helicopters with some synthetic aperture radar scanners borrowed from the Ice Patrol; by 1989 she was down to one short-range helicopter and a pair of binoculars.

In 1990, the *Polar 8* project was scrapped. The Conservatives had won the 1988 election, but British Columbia had voted NDP.

Captain Johns did, however, have an ace in the hole. The Coast Guard owned a larger, Polar-class icebreaker; the *Louis S. St. Laurent* had a 15,000-ton displacement and five steam turbines capable of delivering 30,000 horsepower to her three propellers, easily enough to break through the polar ice pack. She had been built by Canadian Vickers, in Montreal, in 1969, when shipbuilders from around the world were coming to Canada to study Canada's icebreaker designs. She had spent her winters since then in the Gulf of St. Lawrence, keeping the supply lines open between Montreal and the rest of the world, and her summers in the Arctic. And she owned Canada's record for farthest north navigation: in 1971 she had steamed into Alert Bay, on the northern tip of Ellesmere Island.

The problem was, she seemed to be in permanent drydock in Halifax. She had gone in in April 1988 for her midlife refit, a lengthy process that was supposed to add 15 years to her life. She was going to

undergo some major renovations. The shape of her hull was to be
changed and 8 metres added to her length. The thickness of her steel
plating was to be increased by a full 2.5 centimetres. A Wartsila air-
bubbler system was to be installed in her bow. A helicopter hangar
was to be built on her afterdeck. The whole interior was to be redone.
Her old steam turbines were to be replaced with five diesel engines
capable of generating nearly 40,000 horsepower, with an efficiency
that would reduce her fuel requirements from 200,000 litres per day
to about 80,000. (When we left Victoria, her enlarged bunkers had
been charged with 47 million litres of marine diesel fuel, enough for
a year and a half at full throttle in heavy ice, should we need it; we
also carried $2.5 million worth of spare parts, everything from toilet
seats to three spare propellers.) Her entire control room was to be up-
graded and reconfigured and computerized. No one knew how long
this was going to take, or what else the contractor might find once the
ship had been taken apart, or how much it was going to cost (the original estimate was $54 million). When the *Louis* went into drydock
in 1988, the *Polar 8* was still on the books, and in high places it was
asked why Canada needed a super-icebreaker like the *Polar 8* and a
patched-up antique like the refitted *Louis*. But then the West Coast
voted NDP, and the *Polar 8* was scrapped, so everything turned out all
right in the end.

Sort of. With a secure commitment to bring the *Louis* up to a first-
class icebreaker, work on her seemed to lag. Obstacles were encoun-
tered, and after rounds of discussions the obstacles were still there.
When the asbestos insulation was removed from the interior of the
hull, extensive rust damage was found that led to a more intense
structural survey. More money flowed from Ottawa to the Coast
Guard offices in Dartmouth, then across the harbor to the Halifax-
Dartmouth Industries Limited shipyard in Halifax. When the *Polar 8*
was nixed, the original estimate for the *Louis* leapfrogged to $82.3 mil-
lion, then to $100 million, then to $150 million. By the time Eddy
flew to Halifax to see the ship that was going to try to take us to the
North Pole, the price tag read $180 million, and all he saw was a raw
steel hull with cables and wires dangling from exposed bulkheads,
and the engines lying in large pieces in a nearby warehouse. "It didn't

look like a ship to me," Eddy said. "It looked like scrap metal."

The Canadian press, of course, had been in love with the *Polar 8*, with the notion of Canada's having the largest and most powerful ice-breaker in the world. Bigger than anything the Americans had, bigger than anything the Russians had. A ship with a name like a block-buster movie. When the *Polar 8* was scrapped, the press felt cheated, and it looked on the poor *Louis* the way someone expecting a pony for Christmas might look on a puppy.

Egged on by the press, the opposition attacked the government of the day over the escalating price tag. The auditor general suggested in his 1989 annual report that even the original estimate may have been higher than the ultimate value of the finished product. Then Trans-port Canada, the department responsible for the Coast Guard, took a beating from the government. The Coast Guard, in its turn, took a beating from Transport Canada, as well as from the press. Why was the Coast Guard wasting public money refurbishing an old rust bucket like the *Louis*? When it was announced that the refitted *Louis*, on her ice-trial mission to the Arctic in September 1993, had success-fully broken her way through M'Clure Sound, thus becoming the first ship in history to actually sail the true Northwest Passage (the *Manhattan* and the *John A. Macdonald* had been trapped in M'Clure Sound and forced to take the more southerly route through Prince of Wales Strait), a columnist in the *Ottawa Citizen* asked, "So what?" When the Coast Guard divulged that the *Louis* was going to take a group of scientists into the High Arctic to study global warming and pollution, the same columnist wrote: "Great news! After only $200 million in repairs, Canada's largest icebreaker and all-purpose money-sink may become a floating hotel at the North Pole."

As a result, Transport Canada took more heat, and the Coast Guard was nearly ordered to scuttle our expedition altogether, three months before the ships were to set sail. The expedition was saved only when it was pointed out to the various ministers responsible, by Eddy Carmack and David Johns among others, that (a) the $180 mil-lion (not $200 million) had already been spent and so the ship might as well be used for something, and (b) a great deal of other govern-ment money had also been spent on the scientific program, in the form of university and institutional funding, and (c) this was a joint Canada-U.S. project, and Canada would look damned silly pulling out at the last minute simply because some newspaper reporter had likened a grueling, two-month expedition through 5,000 kilometres of unknown polar ice to "a summer vacation" at "the world's most northerly summer camp." The newly elected Liberal government, unlike its Conservative predecessor, did not enjoy the prospect of appearing damned silly in the eyes of its American counterpart.

So the *Louis* was readied for the expedition. "During our ice trials in M'Clure Sound last year," Captain Grandy told me, "we picked the biggest, thickest floe we could find and sailed right through the middle of it. Our intention was to keep going until the ice stopped us, and then see how thick the ice was at that point. Well, the ice didn't stop us. We just kept right on going until we were out of the floe. Later, we drilled a hole through the thickest part of the floe and found that it was 7 metres thick. So, if the ice is not under too much pressure, there's not a lot that will stop us."

The American ship was more problematic. The *Polar Sea* had been designed in the mid-1970s for Antarctic duty, which meant cruising long distances through open water until hitting an ice face consisting of 5 metres of that hard first-year ice Barry Lopez had described, and crashing through that for a day or two until the ship reached Fort McMurdo, on the Antarctic continent. She was originally designed to be a nuclear-powered icebreaker, the first in the so-called free world. As it turned out, the U.S. Navy, which had obtained plans for such a ship from the Soviets, refused to release them to the U.S. Coast Guard. The *Polar Sea* was equipped instead with nine diesel engines, capable of only about 18,000 horsepower, and a secondary engine system of three gas turbines that could provide an additional 60,000 horsepower to the propeller shafts. The idea was that the diesels would get her to the Antarctic ice edge, and then the turbines would kick in for the short-term bashing through the ice. She could not use the turbines for more than a few days, because they used too much fuel: up to 375,000 litres a day at full throttle. The diesel engines at full power would use up her fuel storage capacity in 38 days; the turbos would guzzle the same amount in 13 days.

This made the *Polar Sea* a dubious candidate for extended service in the Arctic Ocean, where what is needed is more like what the *Louis* had: constant thrust of about 30,000 horsepower, suitable for plowing day after day through 2 or 3 metres of hard multi-year ice, with relatively good fuel efficiency. In heavy ice with all five engines running, the *Louis* consumed only about 80,000 litres of fuel a day. (I say "only"; that still comes to about 55 litres a minute.) In the Arctic, where (after ice) fuel imposes the most serious limitation on a ship's range, the *Louis* could carry enough to get her to the Pole and back nine times. Whether the *Polar Sea* could make it even once was unknown. It depended on the ice. If it was thick and recalcitrant, or if it broke one of the *Polar Sea*'s propellers, then the American ship would run into fuel problems. Captain Brigham began to express his worries about running out of fuel the day we entered the ice, and he never really took his eye off the fuel gauge. The *Louis* could afford to be gallant; the *Polar Sea* had to dance through the ice more cautiously.

Captain Johns had mentioned this difference before the trip. He saw it as a good thing. "I think the two ships are perfectly matched," he told me. "The *Polar Sea*, with her diesel engines running, can lead the way in the early stages, before she gets into heavy ice, and the *Louis* can follow in her wake and thus save a lot of fuel. Then, when the ice gets heavier, the *Louis* can take over the lead with her greater horsepower, and the *Polar Sea* can fall behind. If the *Louis* gets into trouble, the *Polar Sea* can come in with her turbines on, get the *Louis* out of the ice, and then drop back behind again, reverting to diesel."

As long as the two ships worked together, there seemed no obstacle to completing the dual mission of science and discovery. No obstacle, that is, but 5,000 kilometres of ice.

The temperature was dropping: the thermometer read −1.6, the first time it had sunk below freezing. I went below to get warmer clothing and passed an open door marked "Gravimeter Room" (the ship's gravimeter, a somewhat dated instrument for detecting ocean-bottom features beneath the ship by measuring variations in the pull of gravity, had been removed during the refit, but the room still bore the old name). Inside, Jim St. John was sitting in front of a computer looking pensive. I went in to talk to him. He said he wasn't pensive, he was exhausted. He and Ron Ritch were monitoring the effects of ice on the icebreakers' hulls to see how the ships' designs held up against the realities of Arctic travel. To this end, the two engineers had spent most of the trip so far attaching 152 tiny sensors to strategic points along the inside of the *Louis's* hull. The sensors were connected to a bank of computers in the gravimeter room. The sensors would record the size of the shock of every chunk of ice that hit the ship's bow, and also the diminution of that shock as it traveled along the hull from bow to midsection. To correlate the size of the shock with the size of the chunk of ice, Jim and Ron had rigged up a video camera on the foredeck, at the end of the port-side yardarm, trained straight down where the ice smashed into the hull. They had been working at it for several days, climbing up the narrow ladder in high winds, duct-taping the square camera to the round mast as the ship rolled with the impacts they were trying to record, climbing back down to check the monitor screen, then climbing back up to change the camera's aim, or focus, or aperture.

They had finally got the camera in place that morning, and now Jim was looking at a screen that portrayed the ice surface and part of the port-side bow. Superimposed over the image was a grid, showing the thickness of the ice as it was being turned on its edge after impact with the ship.

"Finally," said Jim, "we can relax in the comfort of our own gravimeter room and watch TV."

A second screen, beside the one showing chunks of ice rolling over in slow motion, graphically displayed the impact each chunk of ice made on the hull sensors. Every time a piece of ice hit the hull in the screen on the left, the screen on the right showed a large blip opening up along a calibrated line; the larger the blip, the greater the force of the impact. In fact there were four calibrated lines, and I could watch the blips travel from line to line as the chunk of ice rolled along our flank from the bow to the port fresh-water tanks, about a third of the way along the ship's length.

"We also have sensors on two of the ship's three propeller shafts,"

Jim said as we watched the blips. "They measure torque and thrust, or the difference between the strain on the shafts in open water and in various kinds of ice." Each of the *Louis's* propeller shafts was a metre thick and weighed 180 tonnes. Each propeller weighed 275 tonnes; one of the *Louis's* three propeller-shaft assemblies weighed more than the *Jeannette* had when fully loaded and charged with coal. "When the ship is in open water," Jim said, "there is a tiny amount of torque in the shaft: that means the shaft actually twists a bit as it turns the screw. When it gets into ice, it twists a bit more. We want to know how much more, and how much more in 1 metre of first-year ice, 2 metres of second-year ice, 3 metres of multi-year ice, and so on."

Jim and Ron would take this information and apply it to hull designs. Oddly enough, it seemed, icebreakers in the recent past had been designed without much knowledge of what happened when ice meets steel.

"There haven't been a whole lot of studies on how a ship is affected by repeated collisions with solid ice," Jim agreed. "The first was done in 1979, on the *Canmar Kigoriak*, an icebreaker owned by Dome Petroleum. But Dome is a private company and they didn't make their data available."

The *Canmar Kigoriak* featured Dome's patented spoon-shaped hull (unique in 1979, ubiquitous now), which rode up onto the ice so the ship would crush it with its weight rather than waste energy cutting through it. The *Louis's* bow was a modified version. Icebreakers used to ram the ice, a tactic that stove in a lot of bows and didn't cut much ice with shipowners. Much of the research for the *Louis's* hull design had been done by Jim on board the *Polar Sea* in 1982 and the *Nathaniel B. Palmer* in 1993.

"We did seven trips altogether," Jim said of his early work on the *Polar Sea*, "and we measured a total of 3,000 impacts — actual collisions with sea ice — in which we recorded the whole event from bow to propeller. The whole thing only lasts about a second, but it has tremendous reverberations throughout the ship, as you can imagine. Some of those chunks of ice can weigh a couple of hundred tonnes. We're working to a point at which we'll be able to tell where the ship's structure will fail, an aspect that Captain Grandy is naturally very interested in. But we're also interested in the larger picture, at getting a base of criteria that we can apply to icebreaker designs in the future."

"Can you tell now where the *Louis* would fail, if it were to fail?"

"Well, it's early days yet," said Jim. One screen had gone blank, or rather had turned a beautiful cerulean blue, the color of multi-year ice in bright sunlight. "But we seem to be taking some pretty hard hits to the shoulder, just where the bow curves around to the flank. That's where I'd put extra bulwarks if I were refitting this ship."

As I watched the blue screen, a line of words began to appear on it; Jim's screen-saver was a quotation from his favorite book, *Smilla's Sense of Snow*, by the Danish novelist Peter Høeg: "Geometry exists as an innate phenomenon in our consciousness. In the external world, a perfectly formed snow crystal would never exist. But in our consciousness lies the glittering and flawless knowledge of perfect ice."

Since the failures of De Long and Nansen to reach the North Pole via Bering Strait at the end of the last century, there had been only one serious attempt before ours. It was a Canadian expedition, planned by Joseph-Elzéar Bernier, one of the more colorful of Canada's many colorful seafarers. Born in L'Islet, Quebec, in 1852 to a seafaring family, Bernier assumed command of his first vessel at 17, and by the time he turned 40 he had captained dozens of oceangoing sailing ships, mostly merchant ships trading between New England and the West Indies. In 1871, while offloading cargo on the Potomac River, he watched Charles Francis Hall's ship, the *Polaris* — a converted U.S. navy tug — being refitted for Hall's now famous attempt at the North Pole through Baffin Bay. Bernier took one look at the *Polaris's* bow and predicted that the ship would be crushed by ice. In fact, Hall made it to 82 degrees 29 minutes N before being caught fast, a record for the time, and although Hall himself died — poisoned by at least one of his crew — the ship survived two more winters, drifting south, until she was abandoned by her surviving crew members on Littleton Island. In the end, Bernier was wrong; it wasn't the *Polaris* that was defeated by the ice, it was the men who sailed her.

Nonetheless, Bernier had been bitten by the polar bug. From then on, he wrote in his autobiography, published in 1939, "my cabin library on shipboard consisted mainly of books on Arctic travel, and the latest Arctic maps were always in my chartroom." This must have caused some confusion in Barbados. Bernier nurtured his plans for a polar expedition for more than 20 years. In 1895, he was appointed governor of the federal penitentiary in Quebec City — an odd position for a man who had spent his life at sea. Even Bernier was surprised. However, "the appointment flattered me," he wrote, "for it is one usually reserved for retiring members of the Legislature and even Ministers of the Crown. . . . The salary was a mere pittance, $900 a year, with board and lodging for the governor and his family. But I saw the position as an opportunity to concentrate on my Arctic studies."

Bernier bought every book and chart he could find about the north, and studied particularly "what was then known about the currents and tides of the Arctic Ocean." Gradually, he evolved a theory "concerning the drift of ice across the Pole, from the neighborhood of Bering Sea to the eastern and western shores of Greenland."

With the help of a prisoner serving a maximum term for forgery, Bernier made a map of the Arctic Ocean as seen from above the North Pole; on it he traced the suspected and known routes of the *Jeannette* fragments and Nansen's ship, the *Fram*, which had emerged from the ice only a few months earlier. The map also showed his own projected route. "I reasoned that the North Pole could be most surely reached by a vessel which allowed herself to be beset in the ice north of Alaska, the drift of the polar currents taking her across the Arctic Ocean towards Greenland." Bernier familiarized himself with Nansen's calculations, or rather miscalculations, and developed a plan to cross over the Pole by picking a more likely spot in which to take the ice. "We would set sail from Vancouver," he said in an address to the Quebec Geographical Society in 1898, the year Nansen's *Farthest North* was published, "on or about the first of June, stopping at Port Clarence, Alaska, to load the remaining provisions for our dogs. We would then follow the coast of Siberia as far as the New Siberian Islands. We would complete exploration of Bennett Island and Sannikof Island, the latter sited by Dr. Nansen, and perhaps other islands in the area. Next, we would study the movement of the ice and, at a given moment, direct our small vessel into the ice as far as possible at a point some three hundred miles east of the spot where the *Fram* entered."

It was a bold plan, and it caught the imagination of the Canadian people, many of whom subscribed to it — by 1902 he had raised about $20,000 from private sources, still far short of the $150,000 he needed but enough to pique the interest of the Canadian government, to which he applied for further funding. At the time, sovereignty in the Canadian North was a matter of national pride. As the prime minister, Sir Wilfrid Laurier, said in the House of Commons in September 1901, "If a son of Canada were to plant the flag of his country at the North Pole, if he were to achieve what so many brave men have struggled to achieve, there is not a Canadian heart that would not beat with pride at the thought of it." However, Laurier also added the usual kicker: "I had no expectation that the sum of $100,000 would be asked from us." Three years later, Bernier finally received permission from the Department of Marine and Fisheries to purchase, for $75,000, the German polar vessel *Gauss*, and to fit her for Arctic sailing. Parliament also voted to allow Bernier to organize a Canadian polar expedition, with an operating budget of $200,000. After nearly 30 years of dreaming, it looked as though he was finally on his way to the North Pole.

Joseph-Elzéar Bernier, captain of the *Arctic*

Bernier traveled to Germany and bought the *Gauss*, which he renamed the *Arctic*, and sailed her from Bremerhaven to Quebec City in April 1904. There she was fitted out for severe Arctic conditions and loaded up with five years' worth of provisions. Bernier's idea was to get the ship as close as possible to 90 degrees N and then, like Nansen, walk from the ship to the Pole. Unlike Nansen, though, who took dogs, Bernier would travel with trained reindeer fitted with special harnesses for carrying supplies. Reindeer were native to the Arctic, after all, and they tasted better than dogs. He also designed special extendable ladders that could be used for crossing leads and fissures in the ice.

By July, the *Arctic* was ready to sail. "My dream," he later wrote, "was about to be realized."

In fact, his dream was about to be dashed. Just before he was set to sail, he received a letter from the minister of marine and fisheries advising him that he was to sail not around the Horn to Vancouver, but rather into the eastern Arctic, up Davis Strait, and into Hudson Bay, where American whaling captains were reportedly selling alcohol to the Inuit. "After very serious consideration," wrote the minister, "the Government has deemed it advisable that permanent stations should be established at different places on shore in these northern parts of the Dominion; and to carry this out in the best and most effective manner, the sole charge of the expedition has been placed under the command of Superintendent Moodie, of the Royal North West Mounted Police." Bernier had been demoted from Arctic explorer, and possibly the first person to reach the North Pole, back to master mariner, escorting a small revenue expedition into water that had been well known since the 17th century. "I will not dwell on the disappointment that overwhelmed me when this turn of events occurred," he wrote 35 years later.

Bernier never did make his attempt to the Pole. He and the *Arctic* spent four years in the Canadian archipelago, claiming islands for Canada and wresting bottles of rum from the hands of adults on the east coast of Hudson Bay. He continued to dream of going farther north, believing that the revenue expedition was only a temporary check on his plans, but the government kept finding him more important things to do.

In September 1904, a few days after receiving his new orders from the government, Bernier received a second letter, this one from the great Fridtjof Nansen himself. Nansen congratulated him on his ambition, and even offered his assistance to the expedition. "I understand," Nansen wrote, "that you have bought the 'Gauss' and intend with her to sail through Bering Strait, and to make a drift across the North Polar Basin. This is the one expedition which I have advocated, and which I have been looking forward for, because in my

opinion such an expedition would give us a material of scientific observations, compared with which those of other arctic expeditions in
the future would be of little importance." He urged Bernier to send
someone to Norway's Central Laboratory for the International Study
of the Sea, of which Nansen was director, in order to be trained in the
most advanced methods of oceanographic research. Then he wished
Bernier success in his great undertaking: "Were it not that life is so
short," Nansen concluded, "and there are also other things to be
done, I would certainly not hesitate to make another drift with the
Fram from the sea north of Bering Strait."

The letter must have saddened Bernier, but it didn't make him
quit. His career as an Arctic explorer had only begun, except that
rather than follow his original dream — the attainment of the North
Pole — he substituted a new purpose, that of discovering and claiming for Canada as many of the Arctic islands as he could.

There was some urgency. Since the turn of the century, Nansen's
former colleague and the captain of the *Fram*, Otto Sverdrup, had
also switched from polar exploration to nationalist fervor, wandering
around in the Canadian archipelago in the *Fram* for four years,
claiming many of the larger islands he explored for Norway, including King Christian, Axel Heiberg, and Ellef Ringnes Islands (the
latter two named after Norwegian brewers who had sponsored his expeditions). Canada, which had been given sovereignty over the Arctic
by Britain in 1880, had had to buy its own islands back by paying Sverdrup $67,000 to offset his expenses.

Bernier made it his mission to ensure that all the islands were in-
disputably Canadian, and he did so at a frenetic pace. In August 1907,
for instance, he noted in his logbook: "We proceeded on our way and
next day landed on Byam Martin island and took possession of it. At
1 P.M. the same day we landed on Melville island on a projection
which we called Arctic Point in honour of our ship. We built a cairn
and left a record of our annexation of Melville island, Eglinton island,
Prince Patrick island and all the islands adjoining, the whole forming
an area of about 24,000 square miles." Not bad for a day's work.

By 1912, Bernier had made 12 Arctic voyages and, in the name
of Canada, annexed every known and unknown mile between the

mainland and the North Pole. If Canada has any claim at all to the North Pole, it isn't a legal one — no one owns the open sea, as Newfoundland fishermen have found out, and the North Pole is definitely open sea. But it may be an important psychological claim, and its validity is due in large part to the efforts of Bernier. His annexation work, he wrote in his autobiography, "shut the door on possible complications between Canada and the United States, or Denmark or any other country which might have taken advantage of Canadian official indifference to step in and establish some sort of rights to territory already ours."

But his dream of making an attempt on the North Pole was never realized. It was no coincidence that our proposed track line, plotted in advance by Eddy, Knut, and Captain Grandy, was almost identical to Bernier's original course, worked out in the meticulous detail usually reserved for false passports by an imprisoned forger for the voyage Bernier never took.

Eddy had a copy of Bernier's biography with him, on a little shelf hanging above his bunk beside Nansen's *Farthest North*, Elisha Kent Kane's 1854 account of his search for Sir John Franklin and the Open Polar Sea, and A. A. Milne's *Winnie-the-Pooh*, all first editions. (He also had a ukulele and a baseball glove.) Making an attempt on the North Pole, I thought, is a very 19th-century thing to do, like collecting butterflies or writing thick books. I pictured Eddy in his cabin as the latitudes clicked by outside the porthole, reading the accounts of previous explorers who had failed.

"These men never made it to the Pole," Eddy said one day while we were drinking Scotch in his cabin. "But their books might."

7

And Bears

There was a tradition that the Pole was a barren waste, with neither beast nor bird, no living organism in the ocean, and where there was nothing for the scientist to do. Those who attempted to penetrate to these parts for scientific work were regarded as madmen.

— IVAN PAPANIN, *Life on an Ice Floe*

A T 74 degrees N we were poised at the edge of the continental shelf. We were 300 nautical miles from the Siberian and the Canadian coasts, and yet we had sailed over water that was never more than 200 metres deep — the widest continental shelf in the world.

But we were well into the ice pack now. Although the fog was as thick as shrouds, when the wind parted it we could see ice stretching in an unbroken field of white around our entire horizon. It was still first-year, about 2 metres thick, and the *Louis* and the *Polar Sea* were cruising through it at a steady 4 knots, but it was thick enough and plentiful enough to walk on. We began to plan field trips onto it, and to see polar bear tracks everywhere. Malcolm Ramsay and Sean Farley were spending time on the bridge, looking for the bears. Polar bears are intensely curious and fearless animals, and there was a good chance that the two bright red ships or, if the fog remained this thick, the engine vibrations would attract their interest.

Malcolm decided to keep up a 24-hour watch by enlisting the aid of the crew on both ships and any of the scientists who cared to participate. On the *Louis*, he made up a binder with forms to fill out if we saw a bear, or bear tracks, or seals, or fish thrown up onto the ice by the ship's bow, or anything else that might be of interest to a vertebrate biologist entering unknown territory. The binder was kept in a wooden rack on the port side of the bridge, and several of us — including the two Inuit hunters, Ikaksak and Oolatita — worked out a loose schedule of four-hour watches.

I had been in polar bear country before. The first time was more than a decade earlier, when Norman Jewison took a bunch of Hollywood actors to Churchill, Manitoba, on the western shore of Hudson Bay, to shoot the movie *Iceman*, about a Neanderthal found frozen in the Arctic ice and brought back to life. I was writing a profile of the community, and for a while the actors were part of the community. I remember spending an afternoon with Danny Glover and some of the film crew, local boys who were trying to scare us with polar bear stories. Polar bears, they said, were the only animals known to stalk and kill for the fun of it. Polar bears could rip the roof off a car. Polar bears could run faster than humans, and never tire, and never turn back. Glover gave a superb impression of a city man scared out of his wits; I gave a fairly unconvincing performance of not being frightened at all. One afternoon I walked onto the ice north of town, out onto Hudson Bay, the hood of my parka pulled over my head and earphones over my ears. I wanted to listen to Handel's "Water Music" as I looked out over the wind-whipped, frozen surface of the bay, out of sight of land, alone with the ice and the music. Handel's delicate, magisterial phrases and the sight of that harsh, unforgiving icescape lulled me into a sensory rapture. Then I remembered the bear stories, and suddenly every block of ice was a bear, every shadow a bear, every shift and play of ice and sunlight a bear. My stomach tightened and I began to sweat. I turned off the cassette player and, a hair's breadth from panic, made my way back into town. I had never felt so vulnerable or so stupid in my life.

It is good to be cautious in bear country. The polar bear (*Ursus maritimus*) is closely related to the grizzly (*Ursus arctos*) and rivals it for surliness and unpredictability. After so many millennia on ice, it has become like ice. A hundred thousand years ago, when many Asian animals moved across Beringia into Alaska, brown bears migrated from northern Russia into the northernmost regions of North America. Some of them kept moving south and became grizzlies, but others stayed in the North and developed an aptitude and an attitude for living on ice. Gradually, their morphology and behavior changed. They turned white, which made it easier for them to hunt: the Inuit call them "the animal that makes no shadow." Actually, polar bear fur isn't white, it's translucent, each individual hair being hollow and without pigment; it only looks white against snow and ice. In the south, where the absence of color is black, polar bears in zoos and polar bear skins tacked onto den walls look yellowish; up here, they are white. They stopped denning, too, except the pregnant females, who den up in November, have their cubs in January, and emerge from the den in March. Male polar bears do not hibernate, but several times a year they, like the females, indulge in what Malcolm calls "walking hibernation": they go into a kind of daze for a few weeks at a time, sleepwalking in erratic patterns and not eating. They developed a pattern of fasting and gorging that is not known in other bears.

They extended their home ranges to nearly a hundred times that of brown bears: a male polar bear will patrol 2,500 square kilometres of Arctic land and sea ice, an area roughly the size of England. Females are usually less given to peregrination, but one female collared in Alaska in 1992 turned up two years later in Greenland, 4,800 kilometres away. And they developed a taste for seals, the most abundant prey on the ice; so much so that polar bears eat almost nothing but seals, and almost nothing but ringed seals (*Phoca hispida*). "It's an axiom among biologists," says Malcolm, "that wherever you find ice you'll find ringed seals, and wherever you find ringed seals you'll find polar bears."

But ringed seals, the smallest and most common of the Arctic's seal species, are not supposed to be found in the High Arctic. When Malcolm announced at a conference in the spring that he was going to the North Pole to look for polar bears, he was told he was wasting his time.

Polar bears, his fellow mammalogists said, stay within traveling distance of land. Two reasons were given for this. Polar bears like to stay where the seals are, and ringed seals stay with first-year ice because they have to keep their breathing holes open, and keeping a breathing hole open in multi-year ice that is 5 metres thick is not feasible. Besides, female polar bears have to go on land to den, and so the males stay nearby to mate when the females emerge from their dens in the spring. "I was told I would see no bears in the Canada Basin, let alone farther north," Malcolm said. "They laughed when I said I thought I'd find an indigenous population of polar bears north of the continental shelf, bears that live entirely on multi-year ice and never go anywhere near land."

But anyone who reads *National Geographic* knows that polar bears have been spotted almost at the North Pole. In 1990, for example, two Norwegians who were trying to ski to the Pole shot a polar bear at 88 degrees N — 770 kilometres from Spitsbergen. Malcolm considered that too far at sea to be part of a land-based bear's normal stomping ground. And he knew that even in multi-year ice there were enough openings in the ice — small polynyas (open areas in the ice pack), leads, even spaces between blocks in pressure ridges — to provide breathing opportunities for seals. His hypothesis was that wherever there are seals there are bears. He wanted to know whether the bears in the High Arctic constituted an actual icebound population that never traveled to land, not even to den and have cubs. "We've seen bears denning on the ice," Malcolm said, "but always close to land, in fact always within sight of land. I've never heard of a bear denning and having cubs this far north. But then, no one has ever got on a ship and come up to look for them before. So we'll see. The more we study bears, it seems the less we know about them."

Maybe it was our increased alertness, or maybe it was knowing we were on the very limit of the known, edging out over the Chukchi Abyssal Plain, a hole in the ocean floor almost 2.5 kilometres deep, but some of the scientists began to get a little nervous about the possibility of encountering bears, especially those who had to go out onto the ice. "Polar bears behave far differently toward people on the sea ice than they do on land," Malcolm said. "On the ice, they are hunting, and people may be among the hunted."

At one of the science meetings held every day after lunch in the *Louis's* boardroom, people expressed concern about the safety of the ice parties. Knut, Caren Garrity, and I had been aboard the *Polarstern* in the Greenland Sea when, as we watched in mute helplessness from the bridge, a polar bear stalked a small team of scientists working on the ice more than a kilometre from the ship. As the scientists knelt on the ice, absorbed in their work, we kept expecting, then urging, then shouting at them to look up, turn around, see the bear, and make a dash to the helicopter, where the pilot sat watching the scientists. Only they didn't look up. The bear, hidden behind hummocks and all but invisible anyway, was within 15 metres of them when the captain in desperation gave a blast on the ship's horn, alerting the scientists to the danger. After that, no field party was allowed on the ice without an armed escort.

Mike Hemeon and Steve Hemphill assured everyone that on this expedition adequate precautions would always be taken. "There's always a 12-gauge shotgun in the helicopter," said Mike, "and whenever any of you are out on the ice, Steve here stands guard with the rifle."

"Where is it kept?" asked one of the scientists.

"In the cargo hold behind the passenger seats," said Steve.

"What about ammunition?"

"It's in a bag under the pilot's seat."

"So in other words, if one of us had to get the shotgun in a hurry, say if you were not around, we'd have to figure out how to get into the cargo hold, then look for the bag of shotgun shells, then . . ."

"I'm always around," said Steve calmly.

"But what if you weren't? I don't even know how to load a shotgun. Can we get some instruction on this, Mike?"

"Sure," said Mike. "We can set up some weapons training sessions whenever you want. Why don't you make a list of all those who want to participate, and I'll talk to the captain about it."

"We can just get a show of hands right now," said Eddy. "Who wants weapons training?"

Everyone in the room except Malcolm and Sean, who knew how to handle firearms, raised a hand. I looked at Malcolm and sensed that he was not happy about the idea of shooting polar bears, and I wondered why he hadn't said anything.

As we filed out of the room, I asked him: "Would a 12-gauge shotgun stop a charging polar bear?"

Malcolm looked a bit uneasy. "Well," he said, "it depends. The blast might scare it away."

Oolatita makes his living hunting polar bears. The Canadian government issues 650 polar bear licenses each year, and Oolat's village, Grise Fjord, gets five of them. Rather than kill the bears themselves and sell the skins to traders from the south, the villagers sell their quota to white hunters, who come up to Grise Fjord, hire local guides to take them out onto the ice, and end up spending a lot more money in the community than the bear pelts alone would bring. Oolat is one of the guides. He owns a team of sled dogs and contracts with white hunters to take them out for a traditional bear hunt. If they don't get a bear the first day, he builds them an igloo and they stay out overnight.

On the ship, he usually kept to himself, working in the carpenter's shop under the forward hold, where he and Ikaksak were building a komatik — an Inuit work sled meant to be pulled by dogs — that we would use to haul scientific equipment over the ice, or else sitting in his cabin. We rarely saw either Oolatita or Ikaksak in the ship's lounge. But after the meeting he joined us for coffee, and he was downright loquacious about bears.

"What's a good thing to know about bears?" someone asked him.

Oolat thought for a second. "They always turn to the left," he replied.

We all agreed that that was a handy thing to know. Then we asked how bears went about killing seals.

"Ringed seals," he said, "sleep on the ground," by which he meant on the ice. "They make dens in the snow, over their breathing holes, and sleep there." When a bear smells a seal through the snow, it rears up on its hind legs and plunges into the den nose first, grabbing the seal's head in its jaws and crushing its skull. But not every breathing hole is a den, and bears have a different way of dealing with seals that have just poked their noses out of the water for a breath of air. "They hit the ice hard, always with their right paw," he said, hitting the dining table and making the coffee cups jump. "The force of the ice crushes the seal's skull. Then the bear sticks his head through the hole, grabs the seal in his jaws, and hauls it up through the hole. Usually the hole is smaller than the seal, but the bear pulls it through anyway, breaking most of the bones in its body as it comes through. I've seen a polar bear squeeze a seal through a hole that was only this big" — he made a circle with his hands about the size of a bread plate — "and I've seen a bear reach down and pull a 400-kilogram walrus out of the water as easily as if it were a seal. Polar bears are very strong.

"One time me and my cousin took the dogs out to go hunting," Oolat went on. "We made camp a long way from Grise Fjord. We stopped and chained the dogs up and made an igloo, and then we went to sleep in our sleeping bags. A female polar bear came along with her two cubs, and the dogs started barking and getting excited, and they pulled their chains loose and ran off after the bear. This could have been very bad for us, because if the dogs had gone we would've had to walk home and it was a long way. But when they took off, they ran on either side of the igloo, with the chain stretched between them, and the chain cut the top off our igloo. We jumped up fast and grabbed the chain, and stopped the dogs from running away. We were lucky that time.

"Another time, I shot a polar bear but didn't kill it. I was on my snowmobile this time, pulling my komatik. The first thing a bear does when it's shot is bite its wound. The next thing it does is look around for something to chase. After this bear finished fighting with his wound, he looked up and saw me. I started up the snowmobile and started getting out of the way, but before I could get going the bear jumped on my komatik. I took out my knife and cut the line, and got away in my snowmobile far enough to get another shot. I killed it this time, but my komatik was completely destroyed. If I had been out with the dogs that time, I would have been on the komatik myself. I was lucky again."

Our bear watch program began producing results. The ice was criss-crossed by bear tracks, all duly noted in Malcolm's log. Three days

after the meeting, one of the *Louis's* officers spotted a female polar bear
and two cubs on the port side of the ship, at 78 degrees 9 minutes N.
A helicopter was in the air within minutes. Sean, strapped to a seat
with the helicopter door flung open, was ready with the tranquilizer
rifle and made a clean dart of all three bears, catching each one in the
fatty part of the neck. "The trick," Malcolm said, "is to give them just
the right amount of tranquilizer for their weight, so that either the
cubs and mother wake up at the same time or else the cubs wake up
first. If the female wakes up and the cubs are still out, she might think
they were dead and walk off and leave them." The other trick is shooting a rifle through the open door of a helicopter flying 45 knots and
hitting the bear exactly in the side of the neck, not in the head, where
the dart might put out an eye, or in the shoulder, which is so muscular
the tranquilizer might not have the desired effect, which is to keep the
bears comatose while Sean and Malcolm stick their hands in the
bears' mouths. The anesthetic they used was called Telazol, a mixture
of the analgesic Tiletamine and a sedative called Zolazapan. "It was
originally developed in Alabama for the control of prisoners," said
Malcolm, "then the company looked around for commercial applica-
tions and found out it was strong enough to knock out a 500-kilogram
polar bear for three hours. I don't know what kind of prisoners they
have in Alabama, but that's plenty strong enough for our purposes."

When the bears were down, Steve landed the helicopter and Malcolm and Sean went to work. They set up a tripod of tall metal poles
over the female. Then they rolled her onto a rope net slung from a set
of scales at the tip of the tripod, jacked her up off the snow, and read
her weight from scales: 186 kilograms.

"Not very heavy for a lactating female," said Malcolm. Females normally weigh around 200 kilograms; when they are lactating, however,
they should weigh closer to 300. Both male and female bears undergo
tremendous fluctuations in weight: in Churchill, Malcolm captured a
female in December that weighed only a little over 100 kilos; when he
recaptured the same female the following August, she weighed almost
400 kilos. Males also lose tremendous weight during their walking
hibernation stages, dropping from over 500 kilos to 200. "We don't
know how they do that," said Malcolm. "That's the equivalent of you

or I gaining and losing more than 100 kilos every year: our systems couldn't take swings like that, our hearts would give out. How do polar bears do it?"

Once the bear had been lowered back onto the ice, Malcolm and Sean took samples for later analysis. They pulled one of the female's vestigial premolars to get an accurate age (teeth form growth rings much as trees do). With a pair of ordinary nail clippers they took a small tip of claw from one of the bear's forepaws. They also extracted samples of blood, adipose tissue, and in this case milk. The blood and claw tip would be analyzed for isotopes of carbon, which would tell them not only the bear's short-term and long-term eating patterns, but also where the food came from.

"The water off the coast of Russia has a different carbon isotope signature from that off Alaska," Malcolm said. "By finding the ratio of carbon 12 to carbon 13 in the bear's blood, we may be able to tell where its food came from. Personally," he added, "I don't think these bears came from either Russia or Alaska. These cubs are only two months old. They couldn't have moved this far north in two months — we're more than 700 kilometres from Wrangel Island. I think these cubs were born on the ice. I think we're looking at an indigenous population here."

Analyzing blood and claw samples for isotopes of carbon is a non-invasive method of determining a bear's diet; Malcolm and Sean have developed the technique over the past few years to replace the old method: stomach content and fecal deposits. "What's in the stomach and what's in the scat may show what the animal has eaten, but it doesn't show what the animal has digested," Malcolm explained. "In fact, it shows what the animal *hasn't* digested. If all you examine is what's left after digestion, you miss what was 100 percent digested." For example, when a polar bear undergoes its two-month feeding frenzy after one of its walking hibernations, it eats almost nothing but the layer of fat on a ringed seal — it peels the skin and fat off the seal like a glove and leaves the muscle and guts behind for the foxes. Virtually no trace of seal fat is found in a bear's scat, or remains long in a bear's stomach. "Besides," Malcolm added, "in order to take a stomach sample, you have to kill the animal. And apart from the ethical implications, when you kill an animal you only get to sample it once. With tissue and blood, you can sample the same animal several times."

Malcolm and Sean have even used isotope analysis on the fossils of extinct species. "We analyzed some great auk bones from Funk Island, off Newfoundland," said Sean, "and discovered that young auks fed on invertebrates and the adults fed on fish. From that we were able to see what animals moved into that niche when the auks became extinct. We wouldn't have been able to find that out any other way."

Global warming is not the only evidence of change in the Arctic habitat. Chemical contaminants from the south are also finding their way north, and are turning up in ever-increasing concentrations in Arctic animals — in fact, global warming may be increasing the speed at which the contaminants reach the higher latitudes. Sampling the bears' fatty tissue and milk was part of a long-term contaminant study that Malcolm and other vertebrate biologists had been conducting since PCBs became part of the Arctic food chain. Until now, however, all tissue and milk samples had come from land-based bears. One of the things Malcolm wanted to know was whether bears from farther north, away from land and the influence of human habitation, would show higher or lower levels of toxic chemicals than their land-based kin. Apart from the contaminant issue, there were several hard-core biological issues involved, having to do with the bears' ability to gorge and fast, a peculiar cycle that requires them to store enormous amounts of fat.

"In order to build up these huge fat reserves," says Malcolm, "they eat only the outside fat layers of the seals they catch. Because organochlorines like PCB are stored in the seals' fat tissues, polar bears' intake of contaminants can be quite high. What happens to the organochlorines in the female's fat when she begins to fast? Are they broken down by the Cytochrome P450 system of the liver as rapidly as the body fat is used up, in which case the contaminant concentrations in the fat remain constant, or are the toxic compounds recycled back into the remaining fat, resulting in ever higher concentrations in the bear's adipose tissue? We don't know."

The question is important when considering the effect of high toxic levels in the cubs. Female polar bears undertake pregnancy and nursing during the fasting stage. Like all bears, they have markedly small cubs relative to the mother's body size — which is why they can give birth in mid-hibernation without waking up. The birth is timed to take place in mid-fast, when the female is feeding herself and her cub on her fat reserves. The milk they provide the cubs is therefore extremely thick and rich ("Think whipping cream," said Malcolm) and is the cubs' sole nourishment for several months.

If that milk is particularly high in organochlorines, then the cubs are exposed to extraordinarily elevated doses of toxic chemicals at a crucial point in their development. "What would be the impact of such high levels on the cub's nervous, endocrinal, and reproductive systems?" Malcolm asked. "We might imagine that the cubs, whose entire organ systems are in a process of rapid growth and development, would be more susceptible to the contaminants than the adults are. But how that susceptibility manifests itself, we don't know."

In studies of seal populations in Svalbard, high levels of PCBs had a negative effect on the seals' ability to reproduce. If the same proves true for polar bears, and if these bears Malcolm and Sean were capturing in the High Arctic showed relatively low levels of contaminants in their adipose tissues, then a permanent, indigenous, ice-based population of polar bears might be the only viable hope for a fully reproductive polar bear population in the future.

Malcolm and Sean tattooed the inside of the female's lip with an instrument that looked like a pair of wide-mouthed pliers dipped in green ink, so that if the bear were ever captured again whoever examined her would know where and when she had been sampled before. Then they fitted the female with a white plastic radio collar. The collar would send signals detectable by satellite, and over the life of the collar's battery, about two and a half months, they'd be able to track the bear's movements by radar. The collar's fastener was made of an easily corroded metal ("Probably from one of my old cars," said Steve) that snaps after a few months of exposure to salt water, allowing the collar to fall off. I told Malcolm that I had once talked to a coyote researcher in Banff who used radio collars on his subjects, and he'd noticed that radio-collared females tended to mate with radio-collared males. He'd wondered if the collars were acting as a kind of sexual preference factor, like thick beaks in Darwin's finches. I asked Malcolm if he saw the same thing in polar bears.

"No," he said. "We put collars only on the females. With male polar bears, their necks are bigger than their heads, and the collars just slide off. But it's an interesting theory."

The last thing we did before leaving the bears was take their temperatures, using a rectal thermometer. This was Steve's job. "It's hard to stand around looking like a pilot while these guys are doing all the interesting stuff," said Steve. "I'm just glad they dart them first."

8

The Sea Beneath Us

It is now established beyond question that a definite change in the arctic climate set in about 1900, that it became astonishingly marked about 1930, and that it is now spreading into sub-arctic and temperate regions. The frigid top of the world is warming up.

— RACHEL CARSON, *The Sea Around Us*

THE *Louis*'s radio room was on the bridge deck aft of the bridge; it and the chart room and the bridge itself made up almost the entire bridge deck, except for a small bathroom and an electronics room where the ship's ham radio operators had set up their equipment. It was the highest deck on the ship, if you didn't count its roof, which for a reason I have been unable to discover is called monkey's island. Monkey's island is not enclosed, and so except on particularly warm days it is rarely frequented in the Arctic. Above monkey's island there was the crow's nest, another mysterious zoological designation — Herman Melville claims that the first crow's nest was built by a Nantucket whaling captain to give his pilot an advantageous spot from which to look for icebergs. I think it more likely the pilot was up there looking for whales, but in any case Melville doesn't explain why it was called a crow's nest. He might have thought it was obvious, but there are plenty of seabirds that make nests every bit as bulky and altitudinous as a crow's. Why not a heron's nest, or an eagle's?

One day I was passing the radio room, on my way to take my polar bear watch on the bridge, when I heard Gord Stoodley sending out a message in Morse code. The dots and dashes reminded me of old western movies and black-and-white newsreels. As I listened to it flow from his fingers out into the ether, I wondered how anyone could make sense of it. Not many still can. Morse code is going the way of the great auk, with e-mail moving in to fill its niche. In most latitudes, bytes are as easy to transmit as a series of dots and dashes. But we were rapidly moving beyond the reach of satellites, so simpler forms of communication assumed greater importance.

The concentration required to send and read Morse produced a look of pure abstraction on Gord's face; when he was working his CW paddles (for "continuous wave"), he took on the remote, ethereal expression of a concert violinist in the middle of a difficult concerto. It was the look of someone withdrawing from the distractions of the outside world.

This time, though, Gord broke his concentration with a sudden jerk of the head and an oath:

"By the Jeez!" he sputtered, then, seeing me by the door, amended it to "By the jumpin'." I could tell he was upset.

"What's up?"

Gord sat shaking his head for a few seconds, then laughed. He was in such a state of shock that he temporarily forgot that all incoming communications had to be reported to the captain before being discussed with anyone else on the ship, including, or perhaps especially, me. "I was just working Resolute," he said. "The operator there, a guy, said he had a message for me an hour or so ago, but then the captain came in and asked me something about the computer so I signaled 'Wait' to Resolute while I talked to the captain. Well, it was an hour before I got back to him. By the jumpin' he was some ticked off! He told me — this is all on CW, eh? — he told me he had a lot of ships to deal with, and did I think he had nothing better to do than wait around for me, and he went on and on like that. Well, I said, I'm not going to let him get away with that, so I told him I had a lot of things to handle too, the computer data transfer, *Inmarsat A, Inmarsat C, Healthsat A*, the ship's radio, the phone ringing, the fax machine going. So then he said okay, he understood, and he sent me the message. Then he signed off with 77s, which means 'All the best.'"

"So what's wrong with that?" I asked.

"Wait. That was all right, but then he signed off again — this time with 88s! I thought, Holy cow, what's going on here? I mean, we all have our own lifestyles, I guess, but it made me feel kind of jumpy, you know?"

"What does 88 mean?"

"Love and kisses!"

Gord didn't speak to Resolute for the rest of that day. At lunch, when I asked him how his love life was doing, he nearly choked on his mashed potatoes.

"Huh," he said. "I only worked Inuvik this morning. The operator there is a really nice woman named Angela. If I'm gonna get 88s from someone, I might as well get them from her."

An ocean is not simply a tank full of water, all of it the same physical and chemical makeup, as some climate modelers would have us believe. Leaving aside its biological component, an ocean is an incredibly complex and fragile layering of different kinds of waters, each layer separated from those above and below by subtle differences in temperature, salinity, and density — something like one of those layered concoctions of different kinds of alcohol, called shooters, with which university students destroy the brain cells that got them into university in the first place. Each layer of water in an ocean has its own unique source, its own distinct temperature and salinity, its own current, its own floral and faunal assemblage, and its own chemical makeup; in other words, its own history and its own character. All these identifying components are called "signatures." Each layer of water has its own signature.

The signatures of the various waters that make up the Atlantic and Pacific Oceans are fairly well known: oceanographers have been dipping into them for the past 70 years or so, and although things change, often dramatically, they tend to change within certain definable parameters, discernible over time. To a physical oceanographer, a change in temperature at the 1,200-metre level in the mid-Atlantic is much like a change in the weather in your backyard: it might come as a surprise, but if you've been checking the temperature in that spot for 70 years, you'll probably be able to fit the new reading into some kind of pattern. If you can't, then at least you know you've got a temperature anomaly that you can plot to see if it fits into some new kind of pattern. If a new pattern emerges over time, you can begin to talk about climate change.

Very little is known about the history and character of the Arctic Ocean. Indeed, only slightly more is known about it than is known about the moon. The first scientific measurements of it were carried out by Fridtjof Nansen and Otto Sverdrup on the *Fram*, during its nearly transpolar drift in 1896, but they were spotty and represented only a single track line — the haphazard and uncharted drift of the ice. Nansen had assumed that the Arctic Ocean in the higher latitudes was shallow — the theory that there was land, or at least a group of islands, at the North Pole was still prevalent (and was until 1926, when it was disproven by Amundsen in the *Norge*) — and so he packed the *Fram* with depth-sounding equipment consisting of only 200 metres of bronze wire and 1,200 metres of double steel cable. When he drilled holes in the ice beside the drifting *Fram* and lowered the weighted lines with a hand winch, he found himself drifting over water deeper than 1,200 metres. He had his crew unravel the steel cable and join the two lengths together, and add the bronze wire to it, which allowed him to measure depths to about 3,000 metres. He still missed a lot, in some places the bottom quarter of the ocean.

To take water samples, he invented a system of glass bottles, later called Nansen bottles, that he attached to his depth-sounding lines. The bottles descended upside down and were tripped upright at certain depths (determined by the trip wire), filled and sealed, and brought back to the surface. Although these were primitive devices, metal versions of them (still called Nansen bottles) remained in use until Niskin bottles came along in the 1960s, and Niskin bottles are essentially just improved, plastic Nansen bottles.

In fact, the line of descent from the voyage of the *Fram* to the voyage of the *Louis* and the *Polar Sea* is a short one indeed: Nansen's scientific officer on the *Fram* was Otto Sverdrup; Sverdrup's son, Harold, became the director of the Scripps Institution of Oceanography, in La Jolla, California, in the 1950s. Jim Swift, who works at Scripps, was in charge of the physical oceanography work on the *Louis*; although he was too young to have been at Scripps during Sverdrup's tenure, his thesis adviser, Knut Aagaard, was not. A lot of Knut's early work in the Arctic, much of Eddy's, and some of Jim's has been either corroborating or disproving theories of Arctic Ocean circulation and stratification first advanced by Nansen. A scientific paper co-authored by Knut and Eddy in 1994, entitled "The Arctic Ocean and Climate: A Perspective," begins with a quotation from an article by Nansen, published in 1902: "It is evident that the oceanographical conditions of the North Polar Basin have much influence upon the climate, and it is equally evident that changes in its conditions of circulation would greatly change the climatic conditions." They go on to say, "A century later, with climate issues having become scientifically respectable and human society concerned about the consequences of global change, we are now seriously beginning to explore whether Nansen's claims have merit."

There was not a lot of difference between what Nansen set out to discover in 1893 and what we were doing almost exactly 100 years later. Even our methods were barely a generation apart. In some cases we were here to take a second look; in most cases, we were here first.

One fogbound morning in what I called the milking shed — the metal building lashed to the *Louis's* starboard boat deck where water samples were taken from the rosette after it was hauled aboard — Jim Swift told me that the first modern paper in physical oceanography had been published in 1909, written by Fridtjof Nansen and Bjorne Helland-Hansen. Helland-Hansen was the oceanographer who introduced the idea that three simple properties — temperature, salinity, and density — were all you needed to know to identify different kinds of sea water, just as anthropologists divide human beings according to hair texture and skin color.

"That was the start of what I call my kind of oceanography," said Jim, as we watched people come into the shed to drain the rosette's Niskin bottles. Jim looked young to be one of the top Arctic ocean scientists in the world; with blond hair, a wide Howdy Doody grin, and black horn-rimmed glasses, he looked like Tom Swift's younger brother. He often went for a long time without saying anything, and then when he spoke I always felt I'd better write it down.

Nansen and Helland-Hansen used what Jim referred to as the "dynamic" method of elucidating ocean currents. Nansen was a good scientist, well grounded in the basics and also innovative and intuitive. And he and Helland-Hansen were right about most of the things they pronounced upon. "But beyond any doubt," said Jim, "they were wrong about the origin of the Greenland Sea deep water — they couldn't know that there are two types of deep water, one from the Norwegian Sea and one from the Greenland Sea." But then, nobody knew that until the mid-1980s, when Jim Swift figured it out. But about most other things — ice drift, ocean circulation, temperature and salinity curves — their findings are still valid, baselines that all subsequent observations in the Arctic, including our own, take into account.

Oceanography is a young science, which is odd when you consider that 70 percent of the earth's surface is ocean, and that a great deal of human history has taken place on, in, or near it. Sir Walter Raleigh, whose nickname was Sir Water, mused poetically on the nature of the ocean-sea, and early Arctic explorers were hardly more practical. They took quasi-scientific measurements primarily because they knew the people who held the purse strings back home no longer believed much in poetical musings.

As Willard Bascom notes in *The Crest of the Wave: Adventures in Oceanography*, American oceanography hardly existed at all until after World War II. There were a few marine biologists at Scripps, a few more at the Woods Hole Oceanographic Institution in Cape Cod, and that was about it. After the war, writes Bascom, "oceanography gathered momentum because it looked like a fun way to spend one's life." Oceanographers in search of a fun way to spend their lives, however, rarely go to the Arctic Ocean; they develop a keen interest in phytoplankton productivity off the coast of California. And even oceanographers genuinely interested in phytoplankton are not very interested in the Arctic, because the Arctic was deemed to be a desert, and marine biologists don't do deserts. So, from an oceanographical, biological standpoint, the Arctic Ocean remained *mare incognitum* until recently.

"There is still a great deal to be investigated," Nansen wrote in *Farthest North* in 1898, "and this can only be done by years of observation, to which end a new drift, like that of the *Fram*, would be invaluable. . . . I hope that such an expedition may be undertaken ere long, and if it goes through Bering Strait and thence northward, or perhaps slightly to the northeast, I shall be very much surprised if observations are not taken which will prove of far greater scope and importance than those made by us."

As it turned out, it was to be almost a century before anyone made that expedition.

Nansen reported that the surface layer of the Arctic Ocean, the top 50 metres or so beneath the ice, was –1.5 degrees C, which is about as cold as water of that salinity — about 33 parts per thousand — can get without freezing. "Under that," he noted, "the water is warmer, up to plus 1 degree Celsius, and more briny, originating as it does from the Atlantic Ocean by way of Novaya Zemlya, then diving under the colder but lighter and less briny water of the polar basin." The warmest water he found was in the 200- to 250-metre range; then it got progressively colder the deeper his thermometers went, until "near the bottom the temperature rose again, though only slightly."

These observations have been substantially verified by subsequent investigations, made for the most part by scientists camped out on the polar ice cap on drift stations and ice islands. The Russians began setting up ice camps on and around the North Pole in 1937 and maintained them into the early 1990s. Observations from these camps, spotty and kept secret by the former Soviet Union, along with data from similar drift stations maintained by the United States and Canada, represent the only information available about the circulation and stratigraphy of the higher reaches of the Arctic Ocean between Nansen's expedition and ours.

The Arctic Ocean is composed of four principal layers of water. The top layer, from the surface down to about 20 metres, known as the Arctic Mixed Layer, is a layer of relatively fresh, cold water that enters the ocean from the huge rivers draining north in Siberia and North America, and mixes with other water, also relatively fresh, from melted ice. Whatever controls the temperature of this layer controls the climate of much of the world.

Next, from 20 to about 200 metres down, is a layer known as the Arctic Halocline, within which salinity increases rapidly from 33 to 34.2 parts per thousand (the Greek *halo* means "salt"). This layer prevents the ocean from mixing top to bottom; it separates the warmer layer beneath it from contacting the ice cover above it. The halocline, Eddy explained, "comes in two flavors: a thin and relatively weak halocline in the eastern Arctic, and a thicker, more robust halocline in the western Arctic. This thicker and stronger layer is due, in part, to the interflowing of relatively fresh water from the Pacific, through Bering Strait."

Third is the Atlantic Layer, which goes from 200 metres down to about 1,000 metres. Atlantic water enters the Arctic through Fram Strait and the Barents Sea, north of Norway. Because a lot of it is water diverted from the Gulf Stream, it is relatively warm (typically above 0.5 degrees c) and fairly salty, about 34.8 parts per thousand in the Canadian Basin, slightly higher (34.9) in the Eurasian Basin.

The fourth and densest layer, the Arctic Deep Water, reaches from the Atlantic Layer all the way to the ocean floor, which in places is more than 4 kilometres down. This is a slow, sluggish, dense layer of moderately cold water that rarely sees the surface: by processes of convection or upward mixing, deep water rises to the surface only about once every 400 to 700 years.

Until recently, this vertical structure in the Arctic Ocean was thought to be constant and stable; any change in the temperature or salinity of any of these layers was considered either too small to be measured or too slow to have any dramatic effect. Pull up water that is −1.5 degrees c with a salinity of 35 parts per thousand, it was thought, and you know that it came from the Deep Water. Every year 14,000 cubic kilometres of sea ice enters the Atlantic Ocean from the Arctic Ocean. Climate modelers use these assumptions to make their predictions about future climate change. "The modelers all tended to dismiss the Arctic Ocean when computing their climate models," Eddy says. "They took their cue from the oceanographers, who thought that the tiny Arctic Ocean was a steady state, never changed, and all they had to do was run a line of stations across the bottom of the Iceland Sea, figure out how much and what kind of water was entering the Atlantic through it, and then assume that the same thing happened year after year, and was always the same temperature and the same salinity and had the same effect on the climate. Well, we're discovering that oceans aren't like that. Nature isn't like that. Heck, *life* isn't like that."

The study of what happens when two kinds of liquid mix — say, fresh water and salt water, as at the estuary of a river pouring into the ocean, or warm water and cold water, as at the edges of the Gulf Stream or the hysterically unruly Humboldt Current — is called fluid dynamics, and it is one of the most difficult disciplines in science. Suppose you have a bucket of clear water, and you pour into it a cupful of the same water, but dyed blue. As the blue water mixes with the clear, certain visible patterns will present themselves: fronds of darker blue, curlicues of lighter blue, all radiating from a nimbus of color. With experience and a large enough computer, fluid dynamics can predict those patterns and tell you how the water in the bucket will be changed when the mixing is done. Things get trickier, however, when you make the cup of water blue and salty; the fronds and curlicues will be different. Make the water in the bucket room-temperature, and the salty blue water 10 degrees warmer, and the patterns will change again. Maybe the computer will show how the pattern will change, and then predict how the changes will affect the overall nature of the mixed water, maybe it won't. How salty will the water be at the bottom of the bucket? What will be the temperature of the water at the top? And what will be the temperature of the thin layer of air lying along the water's surface? The mind boggles at these questions. So do computers. You begin to appreciate why Einstein said that coming up with the Theory of Relativity was child's play compared with understanding fluid dynamics.

As we are learning, Helland-Hansen's three elements — salinity, temperature, density — change all the time. Nature is about change, but it prefers slow change. Rapid changes occur, but they are usually associated with traumas — the sudden extinction of dinosaurs, the onset of an ice age. Who is to say that water flowing out of the Gulf of St. Lawrence is always going to be the same temperature, year after year, when it meets the cold Labrador Current? Or that the Labrador Current is going to be of a constant salinity? The Humboldt Current, which is supposed to dive under warmer water descending from the equatorial regions off the coast of Chile about every seven years, creating the El Niño effect, decided a few years ago to do so about once a month, for no reason anyone has been able to detect, with disastrous results as far away as a flooded-out farm in Minnesota. Nature abhors rapid change.

In the past two centuries, rapid changes have been caused, or at least accelerated, by human activity. It has been estimated that there is not a bucketful of water in the Great Lakes that has not passed through the cooling system of a nuclear facility or hydroelectric plant at least once. Using water to cool engines or reactors warms up that water. What effect will even minutely warmer fresh-water temperatures in the St. Lawrence have on fluid dynamics when the river meets the Labrador Current? No one knows. And what effect will that change have on, say, cod stocks (since the meeting of the two currents, one from the Great Lakes and the other from the Labrador coast, takes place in the vicinity of the Grand Banks)? No one knows.

"The same thing happens on the West Coast," said Eddy. "The salinity of the water entering the Arctic Ocean through Bering Strait is determined in large part by the amount of fresh water entering the North Pacific from rivers along the West Coast, and for centuries, for millennia, that volume of fresh water has been pretty constant. But in the past two or three decades we have built so many dams on those rivers that the fresh-water component getting into the Pacific has been drastically altered, either in the quantity of water reaching the sea or in the time of year it gets there. Many people don't realize that the timing of runoff is as important as the amount. Take the Columbia River, with its multitude of hydroelectric dams. This river used to feed the North Pacific with vast amounts of snowmelt in late spring and early summer. Natural cycles — fish spawning, insect larva hatching — were tuned to this event. Now we store Columbia River water in the spring and release it in winter, when hydroelectric power is most needed. Good for us, bad for nature. When we disrupt the water cycle, we are tampering with an entire ecology."

Curt Ebbysmeyer, an oceanographer in Seattle and Eddy's old lab instructor, has done remarkable work sorting out natural and unnatural changes in the Pacific. Not only do fish movement and insect cycles differ in winter, he found, but so do ocean currents and wind patterns: the whole ocean is disrupted by those power dams. "Perhaps the worst example," said Eddy, "and the one I'm most familiar with because I grew up in Arizona, is the Colorado River. At the turn of the century, 100 percent of its water flowed into the Pacific Ocean. Today, 64 percent of it is diverted for irrigation in California, 23 percent evaporates in the reservoirs, and only 13 percent gets into the ocean. That means that the coastal waters, at least, receive a lot less fresh water, and therefore are a lot more salty, than used to be the case. What effect does that have on the Arctic Ocean? We don't know. What effect will it have on our climate? We don't know. But we keep building dams.

"At some point in history," he said, "we took water away from nature and handed it over to the engineers. We stopped thinking of water as a living substance and started thinking of it as something that had to be managed, contained, diverted, even sold. Why did we do that? When did we stop thinking of water as someone's home? Why didn't we stop to think that all forms of life, in streams and lakes and oceans, have had tens of millions of years of evolution to get used to cycles, to patterns, to the way the furniture was arranged? These things can adapt to slow change, but we just went in and moved the furniture. Whenever that happened, I think now it's time to take water away from the old-guard engineers and give it back to the biologists."

We have altered the salinity of the water entering the Arctic Ocean. We have also changed its temperature and density. Human actions in the south have affected these changes rapidly, within a matter of a few generations. There is no knowing how the Arctic Ocean will respond to them.

The Arctic Ocean is a fusion of two giant ocean basins, like one of those double kitchen sinks with a stainless steel divider between the two separate — and separately drained — basins. The divider in the Arctic Ocean is the Lomonosov Ridge, which cuts the ocean bed in half along a line from the Laptev Sea toward Ellesmere Island and Greenland, running directly under the North Pole. Each of these basins is divided into smaller basins. The Canadian Basin is separated into the Makarov Basin and the Canada Basin by the Alpha-Mendeleyev Ridge. Water from the Atlantic Ocean entering the Arctic through the Barents Sea and Fram Strait flows in a counterclockwise direction around the Eurasian Basin. When it hits the Lomonosov Ridge, it branches into two main channels; one channel flows over the Lomonosov Ridge into the Canadian Basin; the other stays in the Eurasian Basin and flows along the ridge toward the Pole. That, at least, was the way the Atlantic Layer was assumed to behave.

In the past few years, when most modelers were predicating global warming on increasing levels of carbon dioxide and methane in the atmosphere, and wringing their hands over ozone depletion in the Arctic, Knut and Eddy have proposed a completely different cause for concern. They have suggested that climate perturbations could also be effected by large-scale movements of water masses deep below the Arctic Ocean's ice-covered surface, movements caused by changes in the temperature, salinity, and density of the water in those layers. Suppose more warm water started entering the Arctic Ocean through the Barents Sea than in the past. Suppose this warm Atlantic water swirled around in the Eurasian Basin as usual but didn't sink as deeply into the bottom layer, and therefore, when it hit the Lomonosov Ridge, more of it sloshed over into the Canadian Basin, warming it up. If that happened, Knut and Eddy suggest, "the basic stratification of the Arctic interior itself could be subject to wide fluctuations affecting, in turn, the Arctic's heat budget." In other words, all hell would break loose.

In 1991, oceanographers aboard the *Oden* noted a subtle difference in the Canadian Basin's signature. They did not find the expected abrupt change in nutrients in the halocline layer, signaling the shift, or front, from Pacific- to Atlantic-derived water between the Eurasian and Canadian Basins. The *Oden* sailed 80 nautical miles past the Pole along the Lomonosov Ridge in search of this front, and never found it. They concluded that the front must move around from year to year. This was not alarming, but it was puzzling.

The alarm didn't come until 1993, when Eddy, Robie Macdonald, Ron Perkin, and Fiona McLaughlin sailed into the southern Arctic Ocean aboard the *Larsen*. It was an unusually warm summer, with lots of open water and thin ice, and the ship was able to penetrate parts of the ocean that Western investigators had never been able to reach before. The scientists ran a transect of stations across the bottom of the Canada Basin into the area between the Chukchi Sea and the Mendeleyev Ridge. They weren't looking for the front — it was not supposed to be anywhere near there — but they found it. "It was like stumbling over a log in a dark forest," Eddy said. It wasn't near the Lomonosov Ridge, it was over the Mendeleyev Ridge — 600 kilometres east of where it ought to have been.

The result of that frontal shift was startling. The Atlantic Layer near the Makarov Basin was warmer than it had ever been. Measurements taken from ice camps over the previous 50 years had found no water east of the Lomonosov Ridge warmer than 0.8 degrees C; Eddy and his colleagues on the *Larsen* found water on the Mendeleyev Ridge that was 1.4 degrees C, more than half a degree warmer. In other words, they found that there was a thin section of water, maybe 50 metres thick, within the Atlantic Layer that was warming up significantly and moving farther out into the Canadian Basin. Fiona, who was collecting contaminant data for her graduate work at the University of Victoria, used traces of the refrigerant chemical Freon to "age" the warm water. "Fiona's calculations," said Eddy, "provided the icing for the cake: not only was the water warmer, it was also much newer. The change had occurred recently, and very, very fast."

They hypothesized that the warming was caused by a sudden influx of warm water entering the Arctic Ocean from the Atlantic. "We didn't know whether the water coming in was warmer than it had been before, or whether there was just a lot more of it than had previously been the case, but we did know that something big and new was happening, and it was changing everything we knew about the Arctic Ocean."

Now, after station 12, as the *Louis* began its slow climb up the east face of the Mendeleyev Ridge, Eddy, Knut, Fiona, and Jim Swift knew what they were looking for. Had the front moved again?

9

Handfuls
of Dust

I saw a creature   wandering the way;
And wonderfully   were her wonders worn.
A wonder on the waves   water turned to bone.

—ANGLO-SAXON RIDDLE

A
S WE MOVED off the continental shelf into thicker ice and over deeper water, the biology team on the *Polar Sea*, led by Pat Wheeler, noticed something rather mystifying. The nutrient level in the top 20 metres of the sea remained high — though not as high as it had been in the Chukchi Sea — but large-celled phytoplankton production was way down. There was small-celled phytoplankton, but it didn't seem to be taking up the nutrients (mostly nitrates). Over the shelf, the water had been full of nutrients and teeming with life. There, both kinds of phytoplankton fed off the nitrates, zooplankton fed off the phytoplankton, copepods fed off the zooplankton, and fish fed off the copepods. All was right with the world.

North of 75 degrees, however, the nitrates didn't seem to be translating into phytoplankton, and primary production lagged. "It might mean that the small-celled phytoplankton is taking up ammonia instead of nitrates," said Pat. "It takes less energy to metabolize ammonia."

But it could also mean that the large-celled phytoplankton wasn't growing, and that could be a problem. Phytoplankton, besides supplying grazing material to the herbivorous zooplankton, produces chlorophyll by photosynthesis. Producing chlorophyll uses up a lot of carbon dioxide, which the phytoplankton gets from the surrounding water. The amount of carbon dioxide in the water at the surface is always equal to the amount of carbon dioxide in the air above the water; as the phytoplankton uses up CO_2 the water replaces it by taking CO_2 out of the air. That is what oceanographers mean when they speak of an ocean as a "net sink" for CO_2: high productivity (of phytoplankton) draws lots of carbon dioxide out of the air, reducing the amount of a major greenhouse gas in the atmosphere. If the ocean doesn't produce phytoplankton, it adds CO_2 to the air, and so contributes to global warming. Currently, phytoplankton consumes about half of the 5 billion tonnes of carbon we spew into the air every year: "Kill off the phyto," said Chris Measures, "and we go a long way toward killing off the planet."

The ocean chemistry lab was on the main deck of the *Louis*, tucked between the ship's infirmary and a small room in which Malcolm kept his biology equipment. Chris was a thin man with thick gray hair and a British accent. He had gone to the University of Hawaii because he was interested in dust particles in the ocean, and a lot of dust blows over the South Pacific from Asia. He had joined this expedition to look for trace metals from dust in Arctic sea water, traces so minute that they were measured in nanomolars, which are one-millionths of a gram per litre of water. The apparatus required to make such minuscule measurements consisted of two photospectrometers, each attached to its own Macintosh PowerBook and fed by several miles of fine plastic tubing connected to vials of chemicals and acids, all of which were fed from a small sample of Arctic Ocean water.

Airborne dust is the chief mechanism for the transfer of minerals from land to the oceans. One of the trace metals Chris was looking for was iron. There is notoriously little iron in sea water, which is odd, considering that iron makes up such a large percentage of the earth's crust. As rock breaks down into dirt and dirt crumbles into dust and dust is blown out over the sea, one would think that a lot of that dust would be made up of elemental iron. No one knows why it isn't, but most ocean chemists agree that it's a problem. All plants, including phytoplankton, need iron to photosynthesize. Chris suspected that his instruments would show an unusually small amount of iron in the water we were traveling through, and that the lack of iron would account for the large-celled phytoplankton's inability to process nitrate.

Several other areas of the world's oceans are low in phytoplankton even though they are rich in nutrients. These high nutrient/low chlorophyll areas, or HNLCs, are found mostly in the subarctic and subantarctic regions of the Pacific, but they also turn up close to the equator: there is a 100-square-kilometre HNLC just south of the Galápagos Islands that has been particularly perplexing to oceanographers. Why, in parts of the sea that are warm and rich in phosphates, nitrates, and carbon — things phytoplankton love to gobble up — was little or no chlorophyll being produced? Sailing into the Galápagos HNLC, say oceanographers, is like coming upon a sun-bathed clearing in a forest, with rich loam under your feet, and finding it completely free of plants. It must be eerie to see such miserly growth so near the Galápagos, which Darwin had found so fertile.

The first person to propose a theory to explain HNLCs was John Martin. In 1966, fresh out of the University of Rhode Island, Martin started working for the Atomic Energy Commission on Project Plowshare, the AEC's hair-raising plan to blast a second Panama Canal through the isthmus, right beside the first one, by means of 150 cleverly placed atomic bombs. Martin's job was to calculate what the radioactive fallout would do to the marine life in the Gulf of Panama. This was three years before the first nuclear tests had been carried out on the Aleutian island of Amchitka, so everything was pretty much guesswork. He measured the stable heavy metals already in the water, calculated the amount of radioactive material 150 bombs could be expected to add to it, and then tried to guess whether the gulf's delicate sea life could stand the increase. He guessed it couldn't. The project was eventually scuttled (though not because of Martin), but by then Martin had taken thousands of phytoplankton samples from the Pacific and checked them for trace metal content. Iron is a trace metal. Later, in 1972, as director of California's Moss Landing Marine Laboratories, he began compiling a database of marine trace metals and correlating it with data from HNLCs: in each case, the correlation between low productivity and low iron was remarkable. Areas in the ocean that refused to produce phytoplankton, he hypothesized, did so because they were too low in iron.

As Chris Measures will tell you, iron in sea water is hard to detect. This is partly because there is so little of it, and partly because the samples are easily contaminated by bits of metal floating around in the water. Iron as a trace metal is not the same as a chunk of iron from, say, the side of an icebreaker, but the instruments don't seem to appreciate the difference: touch a tube or a rubber glove with a piece of iron and you could quadruple the iron reading from your sample. Chris wasn't as paranoid about contaminated samples as John Martin was — Martin wouldn't use a plastic bottle until it had been rinsed in hydrochloric acid for at least ten years — but he was acutely conscious that the *Louis* was made of steel, which contains iron, some of which is always leaching out into the sea water around it, and he did wear surgical gloves and tried to keep them away from contact with metal drawers and stools. Even parts of the rosette, which collected the water samples, were made of steel. At the beginning of the expedition, Chris's computer was showing spikes of iron in some samples so high he thought there was something wrong with his program until he discovered that the springs holding down the lids of the Niskin bottles were rusting from exposure to salt water. He took out each spring, polished it, and coated it with epoxy ("I hope no one's sampling for formaldehyde," he said nervously). Even so, he expected some contamination. "I'll just have to isolate it and adjust for it. I mean, that's the reality of doing fieldwork, isn't it."

Martin's insistence on rigorous field conditions and his ability to extract meaningful data from his samples gave his iron-deficiency theory credibility among some otherwise skeptical scientists. He managed to convince a few of them that the reason phytoplankton in the HNLCs was not making chlorophyll was lack of iron. He demonstrated this by taking a beaker of water from an HNLC, adding iron to it, and watching the phytoplankton in the beaker bloom like a forest of trilliums on a warm spring day. Martin then devised what amounted to a multimillion-dollar, science-fiction-like experiment: he wanted to dump 300,000 tonnes of iron into an HNLC in the Pacific Ocean, using crop dusters flying from an aircraft carrier or else spraying it off the back of a supertanker. He figured it would take a few years to seed the entire ocean surrounding Antarctica, but he was convinced that the result would be a massive blooming of phytoplankton that would not only vastly increase the southern ocean's carbon dioxide intake but would also attract more diatoms and krill to the area to feed on the phytoplankton, then seals and penguins to feed on the krill, and so on up the food chain.

At the time, Chris was one of the skeptics. Like a lot of oceanographers, he didn't believe that iron deficiency was a serious enough illness to cause the collapse of the earth's entire nutrient uptake system, or that seeding the ocean with iron would reverse 200 years of fossil fuel burning. It seemed too simple.

But enough other people did take Martin's work seriously and decided to give the experiment a try. In October and November 1993, a group of researchers — not including Martin himself, who had died of cancer the year before and never saw his theory put to the ultimate test — pumped huge volumes of ferrous ammonium sulfate into the Galápagos HNLC from the back of a U.S. Coast Guard vessel, using the cutter's prop-wash as a kind of gigantic Mixmaster. The program, called IRONEX '93, produced mixed results. "Iron turns the ocean greener," reported the team. "Almost immediately after iron addition, physiological indicators of phytoplankton photosynthesis responded positively. After two days, phytoplankton concentrations had tripled." As far as the scientists were concerned, IRONEX '93 was a success and John Martin's soul could rest in peace: it proved that adding iron to areas that are nutrient-rich but phytoplankton-poor caused the phytoplankton to bloom like nobody's business.

Other scientists were not convinced. "They got a huge increase in the standing crop of various species of plankton," said Chris, "which is what Martin had predicted. But, curiously, there was no drawdown of nutrients." In other words, the phyto bloomed, but it did so without feeding on the other nutrients in the water — neither the phosphates nor the nitrates nor even the ammonia — which meant that the sudden growth spurt probably wouldn't have been sustained for long. As it was, the experiment lasted only six days: shortly after the patch was seeded, a huge lake of low-salinity sea water moved over the experimental site and forced the seeded surface water down to a depth of 30 metres, where phytoplankton growth stops anyway.

Then Chris attended a conference of oceanographers in Bermuda, and became a believer. One of the speakers, he said, pointed out that in soils, plant competition for iron was fierce, a life-and-death struggle. Some plants have even developed organic compounds released through their root systems that have no other purpose than to trap iron. Others have ways of disguising the iron they've trapped so that other plants' root systems don't recognize it. And this is in soil, where the iron content is reasonably high. "Well," Chris said, "I realized that if soil-based plants are willing to expend that much energy to obtain and protect their iron supply, then they must really need iron. And if it's that important in soil, imagine how important it must be in sea water, where there is so little of it."

Sea water gets its iron either from the air, in the form of dust particles, or from the ocean bottom, through upwelling, when large plumes of deep water and sediment rise to the surface. Before Martin it was thought that dust and upwelling accounted for about equal amounts of iron in the water. Martin's improved measuring techniques gave a different picture: he calculated that dust provided 95 percent of the ocean's iron content. Chris was suspicious of both estimates, and had come up with his own method of testing the relative inputs.

"I'm looking at two trace metals simultaneously," he explained, "aluminium and iron." Chris is from England and says "aluminium" instead of "aluminum." "Aluminium is as rare and as insoluble in the ocean as iron is, and gets in the same way. But aluminium is more plentiful near the surface than deep down, whereas iron has a deep-water maximum. So I measure concentrations of both metals at the top; if upwelling is the mechanism, then there will be more iron in my samples than aluminium, and if dust is the mechanism, then there'll be more aluminium. It won't give us a final number, of course, but it will be a start."

Privately, Chris was pumping for dust. "I think dust is the key," he said. He believed dust would help explain a puzzling fact about the ice ages, for example. By examining ice core samples from the Greenland Plateau, Martin had determined that ice ages were relatively arid periods when atmospheric CO_2 levels were low — around 200 ppm. Chris reckoned that high aridity would result in greater amounts of dust, and therefore higher iron levels in the ocean. High iron content would allow phytoplankton to bloom, CO_2 would be removed from the water by the phytoplankton, and CO_2 would then be taken from the air to restore air-ocean equilibrium. The result would be lowered levels of atmospheric carbon dioxide, which would lead to global cooling, and so on in a downward spiral to a prolonged ice age. "The question," he said, "is whether it gets colder because the CO_2 levels are low, or do the CO_2 levels drop because it gets colder." The answer has implications now, during a period of global warming. Is the earth's atmosphere warming because of increased CO_2 levels, or are the high CO_2 levels caused by rising temperatures?

Chris's own guess is that the earth is warming because of increased CO_2 levels, and that high levels of CO_2 are causing low iron in sea water. Low iron leads to even higher levels of CO_2. "With our heavy reliance on technology," he said, "we have the potential to trigger vast changes in the global climate structure without really being aware of what we're doing. We don't even know what those changes are — which is one reason we've come up here. But we do know that changes are taking place at a rate never before seen in history. To me, that suggests that the changes are due to human activity. I don't see how we can escape that conclusion."

If dust is the primary source of iron in the ocean, however, how can iron get into the Arctic Ocean? Wouldn't the ice cap act as a giant dustcover? Wouldn't airborne dust settle on the ice and be carried out of the Arctic and into the Greenland Sea and the North Atlantic? Would that explain the low productivity in the so-called Arctic desert? Is the Arctic a vast HNLC?

"Do you read mystery novels?" Chris asked me as he made a tiny adjustment to one of the photospectrometers. I said I had brought along three or four for the trip. "Then you know how some mystery novels are structured so that you know what the final result of an action is at the start, and the rest of the novel is about someone else trying to figure out what happened. That's known as an inverse problem." He looked up and smiled. "We're forensic scientists up here, and global warming is an inverse problem."

On the 3rd of August, at 9 o'clock in the morning, we were stopped at 78 degrees N, 177 degrees W, about halfway up the Mendeleyev Ridge. The ocean below was more than 1,000 metres deep. We were completing science station 14, and while the rosette crew slowly lowered equipment into the Deep Water layer, the two helicopters belonging to the *Louis* took off. One was taking Malcolm and Sean on a second bear-darting mission: an adult polar bear had been spotted by the ice observer on the *Polar Sea*. The second was doing an ice reconnaissance flight. On board were Captain Brigham, Fred Kodz, the *Louis*'s ice observer, and me.

We were looking for leads of open water between the floes. Although the ice was relatively thin, the floes were pressing more closely together, indicating increased pressure in the pack. The wind, at 44 knots and blowing from the north, was pushing the ice down from higher up in the field, driving it against more slowly moving ice farther south. Ice under pressure creates problems for an icebreaker, because once the ship clears a path through it, the path closes in behind, hitting the propellers. If the ice is hard enough and tight enough, the ship stops moving.

"Actually," said Fred, once we had adjusted our headsets, "only about one-tenth of this is first-year. There's lots of second-year ice out there as well."

"How can you tell?" I asked.

"Those larger floes are second-year. They're a grayer blue color, and there are quite a few melt ponds on it. First-year ice is lighter in color, and the floes show no puddles and have clean lines. After a year or more of bashing around, second-year floes have messy edges, because snow and ice have piled up around the perimeter. It looks used. Multi-year, I don't see any of that here, but it has a really bright blue color."

"Killer blue," I said.

"That's what we call it," said Fred. Both he and Captain Brigham were writing their observations on clipboards. "Ridges look higher than 3 metres," said Fred, "but we'll come back at a lower altitude to get a better angle. All the ridges here seem to be oriented east-west, across the ships' path, which isn't great news. Means we'll have to go through them instead of around them. Visibility improving, though. Three to five miles."

We were flying at 78 degrees 10 minutes N, about 12 nautical miles ahead of the ships. Over the radio we heard Steve in the other helicopter reporting to the bridge that the polar bear had been darted and he was landing on a large second-year floe beside it. The ice below us was getting tougher with every mile from the ships, less and less first-year, more and more second-year, and even some multi-year beginning to show up, sometimes contained in larger, lighter-colored floes, like emeralds embedded in slabs of quartz.

"Still lots of pressure," said Fred. "Some of these ridges are pretty mean." There appeared to be a lot of rafting as well, two or three floes piled on top of one another, another sign of pressure.

"Looks like pretty tough sledding," said Captain Brigham, not happily.

Sometime during the next day or two, Art Grantz, searching for his transform fault line, wanted to run a 20-nautical-mile seismic tow through the area over which we were now flying. Somewhere below this ice, he hoped to find Charlie.

A seismic tow is a difficult operation to perform in ice. It involves dragging a large and delicate array of equipment from the stern of the ship. The equipment consists of a 1,270-kilogram steel ball that acts as a weight, keeping a triangular array of huge air guns, or boomers, at a constant depth of 33 metres. Behind the boomer array, at the end of a 180-metre umbilical cord, is a series of 12 hydrophones. When the boomers go off, the shock waves travel down to the ocean floor and hit bedrock; echoes bounce back up to the surface and are picked up by the hydrophones. Because the echoing properties of different kinds of rock are known, computers tell the geologists what kind of rock is sending up the echoes.

To carry out this procedure, the *Polar Sea* would have to tow the seismic equipment in a straight line at constant speed, without stopping or reversing or deviating course, through 20 nautical miles of very difficult ice. That meant that the *Louis* would have to run interference, breaking ice ahead of the *Polar Sea* in a straight line, bashing through anything it came upon — no going around thick floes, no end runs around pressure ridges. The seismic equipment was designed for use in open water, and although it had been towed through ice before, the ice conditions had to be favorable. Even then it was hard on a ship.

We found a few leads heading in the right direction, and for the rest of that day we angled east through ice that seemed determined to force us to go west. Art's target was the Chukchi Gap, a valley on the ocean floor that joined the Chukchi Abyssal Plain to the Mendeleyev Abyssal Plain and, eventually, the Canada Basin itself. If his theory about sea-floor spreading was correct, a seismic tow along the gap would reveal bedrock beneath the sediment identical to the rock he had found at the northern edge of the Canadian archipelago. If you closed the fan, the Chukchi Gap would be filled almost perfectly by Ellef Ringnes Island (just as the bulk of Argentina would fit into the west coast of Africa if you pushed it across the Atlantic Ocean). Art wanted to see if the continental shelf here was a mate to the shelf around Ellef Ringnes. If it was, Charlie would be around here somewhere, too.

But the ice decided not to let us through. At 4:30 the next morning, after the *Polar Sea* had been banging against a pressure ridge for an hour and a half without making headway, Art called Knut and told him he had given up on the idea of a seismic tow and would settle for a piston core right there. It was a tough decision for Art. A piston core would give him a 10-metre tube of sediment going back thousands, maybe millions, of years, but only a seismic tow would tell him what was under the sediment, which could be as much as a kilometre thick; to find out what had taken place millions of years before the sediment was deposited, he'd need to deploy his boomers.

At the science meeting on the *Louis* that afternoon, while both ships were stopped for the piston core, Knut informed us that the ice-breakers would be sticking close together for the next few days. "We're going to be breaking through some very, very difficult ice conditions," he said, "and there is some surety from sharing the risks." In particular, the long hours of bashing through tough ice would be hard on the ships' engines: with one ship leading and the other following in its wake, the second ship would be able to get by using only three of its five engines, saving fuel and allowing the engine-room crew to service the two idle engines.

"Everyone has been working hard and well," said Knut, "except for the marine geology and geophysics team aboard the *Polar Sea*. So far, they have not been able to do much work. They've done some piston cores, but the inability to run seismic lines has been hard on them." Then Knut outlined the plan for the next day or so. We were still going to try to run a short seismic leg over the Chukchi Gap, ice permitting, then go north to latitude 80 to do station 18. "From there," he said, "with any luck at all, we'll follow up the basin, with the *Sea* hauling a seismic tow to station 19. Any questions?"

There were none. This extra jaunt to do a seismic tow over the Chukchi Gap was unscheduled, and everyone was thinking how they could best take advantage of it. Most could use the extra time to analyze the samples they'd taken so far. The note of tension in Knut's voice was disconcerting. It was a good time to hunker down.

After the meeting, I found Caren in the mess studying what looked like a glossy color photograph of a Mondrian painting. It was a satellite image of our section of the Arctic Ocean, with each different ice type showing up in its own distinct color.

"It's from the TeraScan computer on the *Polar Sea*," said Caren. "It shows what the ice is like between us and station 19, along the route that Art wants to take."

TeraScan was an experimental computer program that downloaded images of ice from two polar orbiting satellites, one operated by the civilian National Oceans and Atmospheric Administration

(NOAA) and the other by the Defense Meteorological Satellite Program (DMSP). The NOAA satellite had been more or less useless because it couldn't detect ice through cloud or fog, and we'd been sailing under cloud and through fog almost constantly since we entered the ice. The second, using a special sensor microwave imager, could see through clouds and fog, and we'd been relying on it for our ice maps.

Our present position was represented on the map by a white blip, and Caren had marked the proposed location of station 19, at 80 degrees, with a white grease pencil. Between us and it was the same ice we had flown over a few hours earlier, and the satellite map confirmed what we had seen from the helicopter. Between the blip and the mark, the map was a huge expanse of ice, all of it now one gigantic floe, colored a deep red by the computer.

"What does red mean?" I asked Caren.

"Multi-year," she said. "Bad news. That chunk of ice is 500 square kilometres in size. It'd take us a week to break through it, and a week to go around it."

"Has the captain seen this?"

"I'm taking it up to him now," she said. "I don't think he'll like what he sees."

Neither will Art, I thought. No ship could do a seismic tow through a floe that big.

The bulletin board in the oceanographic lab was the center of most of the scientific activity on the *Louis*. The lab was to Eddy and the other scientists what the bridge was to the captain and his officers, and the bulletin board was its chart table. There was always someone working in the lab, 24 hours a day — Fiona McLaughlin, Louise Adamson, and Mike Hingston checking samples of sea water for CFC contamination; Janet Barwell-Clarke looking for organochlorines; Frank Zemlyak running samples for carbon dioxide levels; Jim Schmitt doing salinity measurements. And almost everyone else passed through at some point in the day. After a plankton haul, it was not unusual to find Malcolm and Sean at the sink, patiently picking copepods and closely related amphipods out of a jar and separating them into piles for a species count. Liisa Jantunen would occasionally poke her head in, asking for help to roll her six 200-litre water tanks from the Challenger pump on the afterdeck into her own lab farther down the passageway, which she shared with Chris.

The lab's bulletin board was therefore a kind of community center, like a bulletin board in a local grocery store. Any message intended for all eyes was usually attached to it. Brenda Ekwurzel reminded everyone not to wear watches with luminous dials in the rosette shed: "The dials are made luminous with tritium," her note explained, "and we're sampling for tritium." Knut posted the latest results of biology sampling on the *Polar Sea*. Jim Swift pinned up a chart of the warm-water anomaly as it spread into the central Canada Basin.

On August 4, just after dinner, Eddy came in to write the science plan for the next few days on the board. We all gathered around to read over his shoulder. It was an unusually long and complicated message:

"Mini-saga," it began: "After a two-or-three-hour CTD rosette cast at station 16, the plan is to continue to 80 degrees N (station 17), where the *Polar Sea* will do a box and piston core (station time is unknown but expected to be about seven hours). A helicopter ice-recon will be made to see if the ice allows passage to stations 18 and 19. (This is where it gets tricky!) If the ice is judged to be light enough to allow a seismic tow, we will go to stations 18 and 19 for box and piston cores and a CTD rosette cast. Station 19 is a biggy for the *Louis*, with expected station time of ten hours. Then the *Louis* will escort the *Sea* (seismically banging away) back to station 17. Meanwhile, Robie and team will have left a Seastar pump at station 17, and we will pick it up before heading northwest to station 20. Is that clear?"

It was clear that the short seismic run over the Chukchi Gap had been called off. Art's next hope was a seismic tow from station 19 back to station 17, a distance of a degree and a half, or 90 nautical miles. At a maximum speed of 5 knots while towing the seismic equipment, such a run would take at least 18 hours, if nothing went wrong. And if the ice were right. If it wasn't, Art's seismic tow would be postponed. Once again, Charlie would elude him.

10

Life on an Unknown Planet

Testing the regions of the ice-bound soul,
Seeking a passage at its northern pole,
Soft'ning the horrors of its wintry sleep,
Melting the surface of that Frozen Deep.

> — CHARLES DICKENS'S prologue to *The Frozen Deep*,
> by Wilkie Collins

CAREN'S TeraScan ice map looked simple enough, a colored-in photograph of the Arctic Ocean taken from a great height (1,000 kilometres or so), but it represented one of the most important breakthroughs in ice detection and forecasting since the Ice Patrol was first set up in 1912, in response to the sinking of the *Titanic*. At first, the Ice Patrol consisted of a few U.S. Coast Guard ships cruising around in the lower Labrador Current, taking note of any large icebergs heading south and radioing their size and the direction of their drift to freighter captains crossing the North Atlantic. Today, the work is done mostly by Canadians; the Ice Centre in Ottawa collects ice information from ice observers, its own Ice Patrol aircraft, and satellites, and relays it to ships in the Atlantic and the Arctic.

Most icebreaker captains cannot receive ice maps directly from satellites; satellite images are downlinked to a receiver in Quebec, from where they are transferred to the Ice Centre, translated into an ice map, and faxed to the ship. All that takes two or three days. In the near Arctic, where ice conditions change hourly, knowing where the ice was three days ago may not seem very useful. But even knowing what the ice conditions are like generally and calculating the direction and speed of its drift can produce a fairly accurate picture of what *should* be out there, and many captains are happy enough to get them. In the High Arctic, however, beyond the reach of Inmarsat, even three-day-old maps are unavailable. This has not been a problem for more than one or two ships every three or four years, but it has limited the shipping season and made exploration hazardous. If the TeraScan program gets up and running, it could revolutionize the way captains anywhere in ice get their information — possibly (we would soon be finding out) all the way to the North Pole. TeraScan can downlink data directly from the satellite. All you need on board is a computer and someone who knows how to read and interpret them, which is where Caren comes in.

Still, satellite imagery is new enough that some captains still don't trust it, won't even look at it. I'd seen Caren arguing with the captain of the *Polarstern* in 1989, trying to tell him that her image obtained from halfway to the moon really did represent the ice conditions a mile ahead of his bow. He was not convinced. Satellite-generated maps are difficult to read: lots of dots and circles colored pixel by pixel by computer, representing who knows what ice concentrations, who knows how long ago. The colors are assigned to each pixel by mathematicians who use algorithms to translate binary data into ice type; lots of room for human or, even worse, computer error.

Remote sensing — the surveillance of ground-level features from satellites orbiting in space — is a new science, and people who have to make instant decisions upon which lives and expensive equipment depend are aware of the gap between research and operations. Perhaps, too, it is unfortunately named: "remote sensing" sounds like someone feeling around in the dark with a long stick. Captain Grandy was one of the few captains who welcomed ice information from any source, the more the better. He had an instinct for ice, one might almost say a passion for it. He talked about ice with Ikaksak. He badgered Gord for more ice maps from Ottawa. He sent Fred out on helicopter ice recons as often as necessary, which was often. And he asked Caren for TeraScan maps whenever they were available. And since remote sensing information was available in Canada largely through the efforts of Caren and her husband, René Ramseier, he knew he was going to the most reliable source. "We've got the best ice people in the world aboard this ship," he told me, "and I intend to make use of them." If there was anyone in the world more passionate about ice than Captain Grandy, it was Caren and René.

I first met René in March 1989, in Gander, Newfoundland, at a series of interdisciplinary ice-science research projects called LIMEX, which sounds like a citrus-based sink cleaner but actually stands for Labrador Ice Margin Experiments, a project put together by the Canadian government. I'd gone especially to meet René, who had been described to me as "the grandfather of remote sensing in Canada." He worked for the Atmospheric Environment Service and was an adjunct professor of ice science at York University in Toronto, where he had organized a small group of ice enthusiasts called the Centre for Research in Experimental Space Science. Within CRESS was a smaller nucleus of graduate students who called themselves the Microwave Group; in 1989, Caren was one of the most enthusiastic among them. She was spending ten months a year at sea on various scientific research ships like the *Polarstern*, trying to finish her doctoral dissertation on the microwave emissions given off by different kinds of ice.

René was experimenting with different scientific ways of reading ice from space. Not just ice movement, which was easy and had been done from aircraft-borne radar since World War II: he was trying to find a way of looking at a satellite image of an ice field and determining from it the internal structure of the ice itself, so that he could point to a particular pixel of ice on the image and say, This is a floe of first-year ice that has rafted up onto a floe of second-year ice, surrounded by broken-up brash ice with a huge chunk of multi-year stuck in the middle of it. At the time, no technology could do that accurately, day or night, summer or winter, cloud cover or no.

In 1989, René believed he was on the verge of perfecting one that could. He was 55, energetic, a man with a mission and government backing. A lucky man. He had a PhD in engineering from the University of Berne and had come to North America in 1959 to do postdoctoral work in cold weather engineering at the University of Chicago. From there he went to work for the U.S. Army in Greenland, helping to build a top-secret under-ice city 100 kilometres inland from Thule Army Base. It was around the time of the Cuban Missile Crisis, when everything was top secret — the more harebrained the scheme, the more top the secret. René and a few other researchers had become experts in the use of ice a building material, having discovered that when ice is chopped up and refrozen it is stronger than ice in its natural state. "We would dig a deep trench in the glacier," he said, "cover it over with some sort of ribbing, spray crushed ice over it, let it refreeze, and the result would be structurally stronger than if we had just tunneled under the original surface."

Ice as a building material had actually been tried before in another top-secret experiment, this one conducted in Canada during World War II. The experiment had been called Project Habakkuk, after one of the later books of the Old Testament, reputedly because the prophet Habakkuk's vision contained the words: "For I will work a work in your days, which ye will not believe, though it be told you." (Habakkuk 1:5) The work in this case was a series of floating platforms of ice that would be positioned across the North Atlantic to be used as landing and refueling stations for Royal Air Force planes hunting German u-boats — aircraft carriers made of ice. The platforms would be either cut from the polar ice pack in the eastern Arctic and towed south, or, if such a thing were feasible, built in huge refrigeration tanks off the coast of Newfoundland, fitted with diesel engines, and driven across the ocean under their own power.

In 1942, C. J. Mackenzie, then acting president of the National Research Council in Ottawa, had received a directive from Winston Churchill ordering the NRC to investigate the potential for such "synthetic aircraft carriers," and if possible to construct a prototype. "I do not, of course, know anything about the physical properties of a lozenge of ice 5,000 feet by 2,000 feet by 100 feet," Churchill told his chiefs of staff, "or how it resists particular stresses, or what would happen to an iceberg of this size in rough Atlantic weather, or how soon it would melt in different waters at different periods of the year." No one knew those things. Mackenzie was ordered to find out.

Mackenzie and the scientists he gathered about him determined to build a floating ice island in Lake Louise, near Banff, Alberta. Perhaps they thought no one in the dining room of the Château Lake Louise would notice a crew of a hundred scientists and technicians constructing a top-secret lozenge of ice measuring 18 by 9 metres, and 6 metres thick, weighing approximately 1,000 tonnes, on the frozen surface of the lake just below the hotel, near the skating rink, beside the changing huts. Or perhaps they thought the Château would be a nice place to warm up after the rigors of the day.

Despite the odds, they actually built the platform, surrounding it with refrigeration coils that kept it intact through the summer of 1943. In the process, they learned several things. They determined that ice mixed with wood pulp was stronger than ice by itself, stronger even than concrete, and so hard that bullets bounced off it and bolts could be driven into it. (It also discouraged tourists from knocking off chips of it for their rye and ginger ale.) The material was called Pykerete, after Geoffrey Pyke, the British inventor who had first proposed the project to Lord Mountbatten. They found that beams of Pykerete, tested at the University of Alberta, had a higher resistance to intra-granular crack initiation than did beams of concrete, or even beams of steel, and remained brittle at temperatures very close to their melting point. They also discovered that the cost of one full-scale ice platform (600 by 90 metres, 60 metres thick) would be around $70 million, slightly more than the cost of a new aircraft carrier made of steel, and that a single hit with a torpedo from a German U-boat would damage it irreparably. The project was scrapped. I wonder if Mackenzie and his colleagues, as they watched their ice platform melting into the cobalt waters of Lake Louise from the broad terrace of the Château, reflected on another passage from the Book of Habakkuk (2:19): "Woe unto him that sayeth unto wood, Awake; to the dumb stone, Arise."

René moved to Canada in 1969 to work for the Inland Waters Directorate, the division of the Department of Energy, Mines and Resources responsible for shipping in the Great Lakes and in Canada's coastal waters. He spent a lot of time on small Coast Guard icebreakers in Lake Ontario, warning tanker captains about ice conditions in the shipping lanes, but he didn't feel his talents were being developed to their potential. He did get interested in designing a hovercraft that would break ice (an intriguing concept, since the whole idea of a hovercraft is not to touch the ice: it worked by injecting air between the ice and the water and sort of exploding it upward), but that was just mental tinkering. The veer toward remote sensing came in 1970, when a tank farm on Section Bay, on the Ungava Peninsula in northern Quebec, ruptured, spilling thousands of litres of diesel fuel into the Arctic Ocean. René was sent up to assess the damage.

"I thought the best thing would be to go up in an Ice Patrol aircraft," he told me in Gander, "and try to map the spill from the air." In those days, the Ice Patrol was using a fairly primitive device called an infrared line scanner, which measured the surface temperature of whatever the aircraft was flying over. A decrease in surface temperature meant you'd moved from open water to ice. René thought that the temperature of clean water would be different from that of water mixed with diesel oil, and so the scanner would give him a clear image of the slick. But the scanner wasn't sophisticated enough to distinguish oil from water; oil had also been pushed under the ice by a slush avalanche, and the scanner couldn't make sense of that, either. "The whole thing bothered me," he said. "I thought it would be a good idea to develop some kind of remote sensor that would detect subtler gradations in surface temperature."

That pretty much describes what he'd been working on ever since. From simple measurements of surface temperatures he moved on to calculating reflective properties. Dark water reflects less sunlight than white ice, for instance — one of the reasons reduced ice cover in the Arctic will increase the rate of global warming — but different kinds of ice also reflect different amounts of sunlight, so detecting and quantifying the differences could enable a computer to distinguish between first-year and second-year ice. That is the principle behind using radar — different ice types reflect radar impulses differently when the impulses are shot down from airplanes or satellites. But radar reflections had their problems: the early versions didn't work through clouds or fog, for instance, and in the Arctic there are clouds or fog 95 percent of the time. Radar scanners on aircraft worked well farther south, for mapping, say, the location of the Soviet Union's wheat fields, or the moisture content of the soil in Alberta, but for detecting subtle differences in ice types René needed something not restricted by weather.

He discovered the virtues of passive microwaves in 1978, when the U.S. Jet Propulsion Lab in Pasadena, California, sent up the world's first nonmilitary satellite. Called *Seasat*, it circled the earth in a near-polar orbit at an altitude of 800 kilometres, sending down images that scientists used to chart ocean currents and ice movement patterns. *Seasat* had the first spaceborne SAR (for synthetic aperture radar, the same system that had been used on aircraft since the war, since replaced by SLAR, for side-looking airborne radar), and everyone in Pasadena oohed and aahed over the images it sent back. (Finding a peacetime use for wartime technology is always satisfying. Consider the ballpoint pen and Teflon.) But *Seasat* also had an experimental passive microwave radiometer on board. The information it sent down was encrypted in a way that hardly made sense to anybody but René. "I began to notice some correlations between brightness temperatures and ice types," he said, "and suddenly I realized that we could distinguish between first-year and multi-year ice using a radiometer."

Passive microwave radiometers measure the naturally emitted wavelengths of objects on the earth's surface. Everything warmer than absolute zero emits passive microwaves — rocks, trees, water, ice, buildings, "you and me," René adds — at different brightness temperatures. Years of patient measuring in the harshest of Arctic conditions have given René and Caren fairly accurate brightness temperatures for different kinds of ice: multi-year ice, for example, has a brightness temperature of .7, first-year ice is .98, open water is .43, and so on. If you know the brightness temperature of snow and you see that brightness temperature on a satellite image, you know you're looking at an expanse of snow and not a field of wheat. But you have to know the brightness temperature of wheat and snow at different actual temperatures, from different angles, and with different water content. It means you have to do a lot of fieldwork, known in the trade as ground truthing. The Inuit really do have 20 different names for snow, but remote sensing scientists have measured hundreds of different microwave frequencies emitted by snow: wet snow, dry snow, snow in September, snow in January, thick snow on ice, thin snow on rock; the variations are endless.

But the advantages of measuring natural emissions rather than beamed radar impulses can potentially outweigh the disadvantages of an infinite multiplicity of wavelengths. One of those advantages is higher resolution, the clearer photograph you get when you're measuring everything continuously, as a radiometer does, rather than the intermittent echoes of radar impulses, as a SAR or SLAR does. You can see smaller objects, objects as small as wavelets on the ocean surface. "I can give you wind speed in the middle of the Atlantic Ocean over hundreds of square kilometres from those wavelets," René said. "The only way we can get mid-ocean wind speed now is from measurements taken on ships, which give it only occasionally and along a narrow track line." The U.S. military shut down *Seasat* after only three months, claiming that its solar panels had stopped transferring power to the main satellite, but there are those who suspect that the Pentagon shut down the experiment because the imagery had such high resolution that an expert could detect the track lines of U.S. submarines. Imagery that clear could have come only from the radiometer.

Telling a computer how to differentiate between new, wet, warm snow on top of first-year ice and year-old, frozen, dry snow on tundra — a difference Ikaksak and Oolatita, for example, can detect at a glance — is a task requiring years of ground truthing and the mathematical acumen of dozens of programmers, and even then it might not be right. On the *Polarstern*, we went out onto ice in the Greenland Sea to measure the water content and brightness temperatures of wet snow and dry snow, old snow and fresh snow, surface snow and subsurface snow; we got dozens of different temperatures, and some of them were the same as the brightness temperature of open water.

How can a computer distinguish between two different substances with identical brightness temperatures? That problem occupied them for years, and was still occupying Caren on this expedition. Most of her fieldwork on the *Louis* involved trying to find a way to distinguish between melt ponds (fresh water) and open sea (salt water): the brightness temperature of both fresh and salt water is .43, and a computer-produced ice map that doesn't distinguish between meltwater on top of a multi-year floe and an expanse of open water isn't much good to a navigator.

It's a task requiring months of kneeling on ice floes with tiny temperature probes and delicate meters, the patient entry of hundreds of precise measurements into sodden notebooks in high winds with frozen fingers, days of transferring data into computers, programmed to translate spindly, splotchy numbers into algorithms that can be worked into xy plots on other computers; then hundreds more hours comparing what the computer says is out there to what is really out there, and trying to figure out why the two don't mesh. When the system is perfected, icebreaker captains will be able to look at a computer screen on the bridge and see exactly what kind of ice they are nudging their ships into, how far it extends, and how fast it is turning into some other kind of ice.

"That day is close at hand," Caren said. "But then, it's been close at hand for 15 years."

Remote sensing of both poles has given us a fairly accurate fix on what is happening with the world's ice, both sea ice and glacier ice. That has been important for global warming researchers, because it allows them to chart change over time. One of the changes it has shown us, for example, is that the world's total ice load is shrinking. That glacier ice is melting is not necessarily a cause for immediate alarm, but it does lead one to consider what the world would be like without it.

Let's first consider what the world was like when there was too much of it. During the Wisconsin Ice Age, the last big one, which lasted 60,000 years and ended 11,000 years ago, the high latitudes from the Pole down to Indianapolis and Stockholm and Glasgow were burdened with a layer of ice that was, on average, more than a kilometre thick. The same was true for the high southern latitudes. We know the extent of the ice from a wide variety of evidence: fossil tree rings, polished bedrock, valleys and fjords gouged out by glaciers inching down like snowplows, huge deposits of till in places where you'd least expect them. It was the German poet Goethe, a serious amateur geologist, who first proposed the notion that erratics and drift boulders in the Swiss Alps and northern Germany had been carried there by glaciers during an ice age, a term he invented. "It is my conjecture," he wrote in 1829, "that all of Europe, at least, passed through an epoch of great cold." Not just Europe: drive past Peggy's Cove in Nova Scotia and you will see enormous stranded boulders standing on the flat granite sea shelf like giant billiard balls on a slate table, rolled there from hundreds of kilometres away and spat out by retreating tongues of ice.

Getting a sense of the volume of the ice cover is trickier. We can map out the area that was covered by ice, we can chart its advances and retreats over consecutive battles with time, but can we determine how thick the ice was? We know that the earth's crust has sprung back several metres since the ice melted, and we can calculate how much weight must have been removed to allow that to happen — the center of Greenland, for example, has been pushed far below sea level by the weight of the ice upon it, and southern Ontario is still rebounding from the release of the ice's weight after the last ice age — but there are variables and unknowns in the equation. Still, thicknesses of up to 2.5 kilometres in the higher latitudes have been bandied about at glaciological conferences.

One measuring device does give us some idea of the magnitude of the ice: during the last ice age, the world's oceans were 100 metres lower than they are today. Since the amount of water on the planet is constant, the water that now makes up the top 100 metres of the world's oceans was somewhere else until 11,000 years ago. The total area of the earth now covered by ocean is 360 million square kilometres; the top 100 metres equals 36 million cubic kilometres, or approximately the same amount as is now found in the Antarctic and

Greenland Ice Sheets. Some of it may have been present in the form of vapor — clouds and fog — but most of it was tied up in ice.

The Wisconsin Ice Age ended with an incidence of global warming; ice core samples from Greenland suggest that during the warming period, the average annual temperature in Greenland rose by about 8 Celsius degrees, which means that the global average temperature probably rose by only 2 or 3 Celsius degrees. If current predictions about our present rate of global warming are right, we'll reach a 3-degree average temperature rise around the middle of the next century, and a 5-degree rise by 2100. Such an increase will easily be enough to rid the planet of all its current ice.

This would raise the world's oceans by more than 75 metres. A world without ice is a world with a lot less land; we're talking *Waterworld*. The melting of the Greenland Ice Sheet alone would raise ocean levels by 3 metres; if the East Antarctic Ice Sheet goes, as some say it is now going, it would increase ocean levels by 30 metres. Vancouver's City Hall is 11 metres above current sea level. Tokyo is 5 metres. Cape Town, South Africa, is just under 12. Los Angeles, at 95 metres, would lose its coastline. Forget New York: La Guardia Airport is 5.5 metres above sea level; Brooklyn is 4.5. Melting vast sheets of ice is a slow process that could take as long as 500 years, but melting even the smaller glaciers on the planet over the next 100 years would have drastic effects on our lives and on those of our children. The Alaskan glaciers, which have been melting steadily for several decades — they already contribute as much fresh water to the Pacific Ocean as the St. Lawrence River does to the Atlantic — would add half a metre to the world ocean level, and if they went so would most of the rest of the world's glaciers. It has been estimated that melting Alpine glaciers in Europe have raised world ocean levels by 5 centimetres in the past 100 years. If they and Alaska's glaciers disappear, they would add a metre and a half to the ocean level by 2100. The Netherlands' massive seawalls were built to withstand ocean fluctuations of only half a metre. Countries such as Bangladesh and parts of Thailand are now less than a metre above sea level. A metre rise in sea level would bring the shoreline of eastern North America inland by about 30 metres.

Low-lying countries are worried, or should be, because scientists have been telling us for at least 20 years that the world's sea level is rising. A *Scientific American* review of Stephen Schneider's *Global Warming* noted that "the rise of sea level is the most probable and the most globally uniform consequence of any warming, and perhaps the most ominous." Schneider put the current rate of sea level rise at 2.5 centimetres per decade, a rate he predicted would accelerate as global warming increases. Two researchers at the University of Colorado announced in 1989 that "the ocean level is rising at a rate of 1 to 2 millimetres per year, but is expected to rise much faster, perhaps a metre or more by the year 2050, if predicted greenhouse warming trends occur." About 65 percent of that rise they attributed to thermal expansion — warm water occupies a greater volume than cold water — but the rest "cannot be accounted for," they said, unless it comes "from an increased ice discharge into the ocean from the ice sheets covering Greenland and Antarctica."

Melting glaciers are adding tremendous amounts of fresh water to the world's oceans, and the shrinking of the Antarctic and Greenland Ice Sheets would be catastrophic. Even if it takes 500 years to raise the sea level 75 metres, 500 years is not such a long time. Columbus sailed to America 505 years ago. Shakespeare was born 430 years before our ship left Victoria. These events are well within our cultural memory; their effects are still with us, and we know where they came from. It is possible to read all the extant works of English literature written between 1100 and 1600 A.D. in a single semester (or at least convince a professor that you have). We cannot dismiss global warming simply because it is happening slowly. Even if the average temperature is rising only half a Celsius degree a decade, by the time my granddaughter is my age the world's average temperature will be 2.5 degrees warmer. At the North Pole, it could be as much as 10 degrees warmer, which means massive melting of what will no longer be the permanent polar ice pack. Ice in the Arctic will be seasonal, as ice on the Great Lakes is today. And there will be no ice on the Great Lakes.

Rising sea level is only the most dramatic consequence of global warming, and significant sea level rise is only a theoretical possibility. The Arctic's ice cap is already floating in the ocean and wouldn't add a millimetre to the sea level if it all melted tomorrow. But what most people who talk about a melting Arctic ice cap forget, and what Eddy and Knut and Jim Swift were realizing during this expedition, is that if the polar sea ice melts it would have a devastating effect on the world ocean's heat dispersal system, and therefore on world climate. And that, too, is already happening.

The world's oceans absorb and distribute vast amounts of heat from the sun. About 70 percent of the solar energy that hits Earth hits water and is either reflected or absorbed. The oceans are the earth's heating and cooling system. They are gigantic heat sinks. The world ocean circulates heat that is absorbed in the tropics (where most of the water is) and carried north, into the North Pacific and North Atlantic regions, where wind blowing over it keeps the landmass in the northern latitudes temperate. As it cools down near the poles, the water is then circulated back south, where it cools off the equatorial landmasses. It is not merely the presence of the oceans that keeps the planet habitable; it is the circulation of those oceans, which ensures the transfer of heat from regions where there is too much of it to regions where there is not enough of it, that makes human life on this planet possible.

One of the three things that make the oceans circulate the way they do is the amount of salt in the water. (The other two are temperature and density, both of which are affected by salt content.) Different layers of water within the ocean are kept separate by differences in their salinity. If the salt content of warm water leaving the Indian Ocean, for example, were slightly lower than it is, it would mix with water in the Atlantic rather than flow through it like a river, and the Indian Ocean's heat would not get circulated north. This would cool down the Gulf Stream and make England's climate more like that of Labrador. If the salt content of water leaving the Arctic Ocean were lower than it is now, cold Arctic water would mix with the warm Gulf Stream coming up from the south, instead of sinking below it. It therefore would not travel south to replace the warm water leaving the Indian Ocean, and would not cool down the tropics. Alter the salt content of a significant part of the world ocean, in other words, and you alter the entire global heat transfer system. You alter the world's climate.

Second-year and multi-year sea ice is fresh water in solid state. When it forms, from frozen sea water, it is salty, but after two years all the salt crystals trapped during freezing have leached out into the ocean, leaving the ice so salt-free that the summer melt ponds on top of it are drinkable. I've tasted them: they are warm and slightly salty, like tears, but they are fresh. When the Arctic's ice drifts from Siberia across the North Pole and down into the North Atlantic, where it eventually melts, it adds fresh water to the Atlantic Ocean, more than the four largest rivers emptying into the Atlantic combined. If the polar ice pack melted, it would add tremendous amounts of fresh water to the Arctic Ocean and take away tremendous amounts from the Greenland Sea. It would drastically alter the salt content of water meeting the Gulf Stream. It would drastically alter the ocean circulation system and the world's climate.

As early as 1986, researchers using remote sensing noticed that the polar ice pack had shrunk by 6 percent in a decade. That is a fairly small shrinkage in area, but what about volume? In 2002, NASA plans to launch a polar orbiting satellite that will monitor both the Arctic and Antarctic ice caps for changes in height as well as in area, and NASA claims that this will enable detection of subtle changes in the overall volume of ice stored at the poles. But it will really be effective only in measuring the volume of *land* ice, the glaciers in Antarctica and Greenland. Sea ice in the Arctic, the so-far-permanent ice cap that is four times the area of Greenland, will not change in height when it melts, because it will melt from the bottom up, not the top down. Investigations by submarines under the Arctic ice in 1989 showed that the thickness of the ice pack had already diminished by as much as 2 metres in some places; Knut Aagaard has recorded similar thinning in sea ice entering the Greenland Sea over the past three years. Two metres represents an almost 40 percent shrinkage. By the time NASA's new satellite sends back data showing no change in the height of the Arctic ice cap, there may be nothing left under the snow but a thin crust of ice, and the entire fresh-water budget of the northern Atlantic Ocean may have changed drastically.

"And that means," as Eddy put it, "we'd be living on a whole different planet."

11

Alphabet Soup

Acid rain, desertification, the greenhouse effect . . . illustrate a
fundamental flaw in the relationship between human society . . .
and that interaction of land and air and water that collectively
produces the conditions we call climate.

    – THOMAS LEVENSON, *Ice Time*

A T OUR FIRST STOP at station 17 we were just a nautical mile shy of 80 degrees N. I went out on the ice with
Darren Tuele and Dave Paton, technicians from IOS
who were working with the contaminants program. They needed
someone to help them push the komatik, which was loaded with
heavy water-pumping equipment. We walked some distance from the
ship to find a big floe, pushing the komatik over narrow leads and high
ridges until we came to a wide expanse of ice, as open and inviting as a
baseball park, albeit with a tumble of ice blocks crossing center field.
The *Louis* and the *Polar Sea* were both hidden from us by a 6-metre
pressure ridge. It was as though the ships had ceased to exist, and nothing but a crust of ice separated us from 2 kilometres of water, straight
down. The sensation produced in us a jocularity we could hardly contain: we ran across the ice, dragging the sled and laughing like children at a toboggan party. The only other sounds were the hissing of

the sled's metal runners, the sighing of wind over the frozen fields of snow, and the haunting cries of gulls coming at us through the fog. It did not occur to any of us that we had not brought along a rifle.

We were going to lower four Seastar pumps into water above the Chukchi Gap. On the sled was $100,000 worth of equipment, including an electric auger with enough add-on bits to drill through 4 metres of ice, a gasoline-powered generator, the Seastar pumps, several lengths of nylon rope, two bright orange marker buoys, one large blue plastic barrel, a 2.5-metre length of spruce, an ice chisel on a 3-metre wooden handle, and an ordinary aluminum snow shovel. The sled was one of two komatiks that Ikaksak and Oolatita had made. It was a modernist version to be sure, consisting of dimensional lumber and yellow polypropylene rope instead of whalebone and walrus sinew, but it moved across the ice on its metal runners like a pat of butter on a warm skillet.

We slid down the side of a pressure ridge onto an expanse of flat ice, then dragged the komatik out to the center of the floe. Another pressure ridge loomed 30 metres ahead of us: we were in a sort of valley between two high, parallel ranges of ice. Darren started up the generator while Dave, who was taller, plugged the auger into it and started the motor. The noise of generator and auger seemed like an affront after the whispering silence of the ice. We needed two large holes about 6 metres apart, each hole at least half a metre in diameter. Having hacked similar holes with an axe through a mere 30 centimetres of lake ice in Ontario, a job that involved half an hour and quantities of sweat that would be dangerous in the Arctic, I appreciated the auger. It made beautiful 15-centimetre holes.

Dave pressed the auger down through the ice. When he reached the depth of the bit he stopped the motor, removed it from the bit end, attached a second bit to the first, reconnected the motor, then started drilling again. A slurry of gray slush accumulated around the hole. He changed bits again, then stopped.

"What's the matter, Dave?" said Darren. "Tired already? Want me to take over?"

"No, but this is one thick hunk of ice you put us on. I'm down 3 metres and still not through."

"Hmm," Darren replied. "This is just second-year ice. We must be on two floes rafted together." If the hole was deeper than the chisel handle, they wouldn't be able to ream it out at the bottom. "We'll have to find another spot."

"Great. At this rate we'll have a golf course in no time."

We moved to a spot farther out where the ice seemed thinner, and tried again. After an hour's drilling and chiseling and drilling again and shoveling, we had two good-sized holes and were ready to lower the four pumps. Seastar pumps are Canadian-made contaminant monitoring devices consisting of a battery, a metal cylinder about the size of a scuba tank, a pump, and a series of filters, all suspended vertically in the water on rope. Depending on the kind of filter, the unit would collect whatever it was told to collect. In this case, Dave and Darren wanted to know about organochlorines, so the filters were a type that would absorb such organochlorines as PCBs and HCHs and let other contaminants pass through. A counter measured the exact amount of water that flowed through the cylinder. "When we get home," Dave said as he lifted the four Seastars from the komatik, "we'll measure how much organochlorine is in the filters, then we'll ask the Seastar how much water it pumped, and we'll be able to calculate a parts-per-million from that. Pretty simple, really. Can't think why these things cost $12,000 apiece."

We tied two pumps to a length of rope and lowered them through one of the holes; then we did the same with the other two pumps, so that the four were suspended at depths of 200, 150, 50, and 10 metres. Darren tied the two lines together and ran them through an ice anchor driven into the ice between the holes, then attached an orange floater to each line in case the floe split where the holes were. Then he tied the plastic barrel to the ice anchor to act as a third float while Dave set the *Louis*'s Raycon beacon up on a tripod beside the whole array. "The Raycon gives off a signal that can be picked up by satellite," said Darren. "If we come back this way after station 19, we should be able to pick up the signal and retrieve the pumps even if the ice has drifted. Unless, of course, the Raycon doesn't work."

"What if we don't come back this way?" I asked.

"We'll have to send a helicopter back. We'll still need the Raycon."

"We had a radar beacon attacked by a polar bear once," Dave said as we repacked the sled. "He knocked it over and it took a hell of long time for the pilot to find the buoys. Anything can happen up here."

Hanging Seastars from the ice is a relatively neat way of deploying them. In open water, they have to be anchored to the bottom in some way, with the upper end of the rope attached to a large buoy that will hold them vertically in the water while they pump away. The buoy sits about 10 metres below the surface. In 2,500 metres of water, you need a strong buoy and a heavy anchor. The line is attached to the anchor by an acoustically operated release mechanism: when researchers on the ship want to retrieve the pumps, they send a sonic signal to the release mechanism, and the buoy obediently bobs to the surface, Seastars attached. "Oceanographers have deployed thousands of these things over the years," said Darren as we hauled the komatik back over a high ridge of tumbled chunks of ice. "They work really well, now that we got the bugs out of them."

"What kind of bugs?" I asked.

"Well, for one thing, the release mechanisms were sometimes triggered by the sound of a ship's propeller, so any time a ship happened by, the buoys would pop to the surface."

"What happens to the anchors?"

"They're left on the bottom," Darren replied. "The ocean floor is littered with them. Lots and lots of train wheels."

"Train wheels?"

"That's what everyone uses. Old train wheels, two of them attached by an axle. Why not? They're cheap, there are lots of them, and they're the right weight. You can't very well haul them up."

I was thinking about Chris Measures and his iron particle measurements. What if one of his deep-water sample bottles were tripped a few centimetres from an old train wheel that had been eroding away on the Arctic seabed for a few dozen years? How many nanomolars of iron would there be in a 1-tonne train wheel?

I had been pushing the komatik from behind, up a steep pressure ridge, and when it reached the top and began sliding down the other side, toward the ships, I hung back, watching Darren and Dave pull it across the neighboring floe. I took off my mitt to wipe my forehead, and dropped the mitt in the snow. When I bent to pick it up, I stopped thinking about train wheels. The mitt had fallen into a perfectly formed, very large, fresh-looking polar bear paw print. I remarked on the clarity of the heel pad, the humanlike spread of the fingers, the unhumanlike distance between the claw marks and the toe pads. Large male. I picked up my mitt and ran like a mad fool after the komatik.

As the ships began working north toward station 18, leaving the Seastar pumps behind, we moved into a bank of heavy fog that kept visibility to a few hundred metres. Malcolm called off the bear watch, but I found Robie Macdonald on the bridge with binoculars, watching the ice chunks turned up by the *Louis*'s prow. We were in the heavy multi-year floe that I had seen on Caren's map, and the huge, greenish blocks slowly assumed the vertical before easing over onto their backs and scraping along the sides of the ship. Jim St. John and Ron Ritch were getting plenty of data points. We turned over an entire Stonehenge about every five minutes.

Robie was fishing. Arctic cod (small members of the family Gadidae) didn't exactly abound in these waters, but occasionally, once every two days or so, the ship's bow would splash one up onto the ice as it plunged through the pack. When that happened, Robie would signal Captain Grandy to stop the ship. Then he would dash below, don his Mustang suit, have himself swung over the side by one of the forward winches, and scoot back along the ice to collect the fish in a Ziploc bag, which he labeled with the date and exact location of his catch.

Robie had realized, after the trip began, that he could incorporate an analysis of fish lipids in his Arctic contamination study. "Fish are actually miniature Seastar pumps," he said. "They are beautiful integrators of contaminants." The fish live on copepods, which accumulate contaminants in their fatty tissues and pass them on to the fish; all Robie had to do was extract the fishes' fat and analyze it. "They don't sort the organochlorines out into different categories for us, like the Seastars do, and we don't know the exact depths they're sampling for us, but boy, do they accumulate parts per million." A single arctic cod could contain levels of organochlorines that are 5 million times higher than the level in the water they're swimming through.

The Arctic isn't an ocean in which you'd expect to find industrial wastes, and yet they are here — another indication, if we needed one, of how busily we have been trashing the planet for the past century and a half. Polychlorinated biphenyls, or PCBs, were introduced in the 1930s as coolants for high-voltage capacitors, those big black boxes that sit on top of telephone poles and hum worrisomely when it rains. Since PCBs can withstand temperatures up to 870 degrees C without deteriorating, they were useful in transferring heat out of electrical transformers into the outside air.

In 1966, a Swedish chemist named Sören Jensen began finding high concentrations of PCBs in odd places: in a fish caught in Stockholm harbor; for example; in an eagle feather from the Swedish archipelago; in hair taken from his wife's head; and, most alarming of all, in the hair of his newborn daughter. Soon scientists were finding environmental PCBs just about everywhere they looked: in porpoises in the Bay of Fundy, in Adélie penguin eggs in Antarctica, in coho salmon from the Great Lakes, in trout from northern Quebec, in cod from the Baltic Sea. Then researchers began to notice that certain chemicals, polyvinyls, for example, which are closely related to PCBs, were causing abnormalities in laboratory animals: neurobehavioral damage, birth defects, respiratory ailments, and so on. At the Carnegie Institution's embryology lab in Baltimore, heart cells from chicken embryos began mysteriously to die when the lab switched from glass to polyvinyl containers. Polyvinyl was also linked to problems in blood transfusion patients in Vietnam and at the Johns Hopkins Hospital in Baltimore; blood kept in plastic containers tended to clot.

Arctic birds ended the conjectural phase of PCB research. In 1969, Helen Hays and Robert Risebrough conducted an investigation into birth defects in common and roseate terns on Great Gull Island, a small, rocky outcrop in Long Island Sound, just downstream from New York City. Despite its name, Great Gull had once been a nesting site for a huge tern colony, but during World War II the U.S. Navy had used it as a lookout post, an occupation that apparently required that the entire island be paved over with asphalt and concrete. The terns disappeared. In 1949, however, the American Museum of Natural History purchased the island and designated it a tern sanctuary, and at the time of Hays and Risebrough's arrival more than 7,000 birds had returned. The researchers found tern nests, but what they discovered in the nests was so disturbing that they knew they had stumbled onto a major ecological disaster. They found tern chicks with four legs. Others with one eye. Chicks with "thalidomide-like" feet and wings. Three-week-old nestlings with no wing feathers. They found one chick with no lower beak, and with its upper beak growing out of its forehead. "There was no doubt," they wrote in Natural History magazine in 1971, "that we were looking at the tip of a potentially disastrous iceberg."

They realized that the birds were being exposed to a chemical "capable of affecting gene reproduction during embryonic cell division," but they didn't know what it was. Some of the defects — thin-shelled eggs, for example — had been linked to DDT in ospreys and bald eagles, but the Great Gull Island terns showed low levels of DDT and its derivative DDE. But they did have high levels of PCBs, which, they knew, had chemical properties similar to DDT. Since PCBs are stored in fat and are thus bioaccumulative, the researchers were not surprised to find high levels in terns, which are near the top of the food chain. They looked upon their terns as the proverbial canaries in the mine shaft: Great Gull Island had served as an early warning post during the war, they wrote, and it was still serving that purpose now: "The terns occupying the approaches to New York City are now inadvertently detecting a newer, more insidious enemy to man: contaminants."

At first there was some mystery as to how so many PCBs were getting loose. Most of them were used in so-called closed systems — refrigeration coils in high-voltage capacitors, cooling mechanisms for fluorescent lights, and commercial air conditioners. Some of these things no doubt leaked, or were dumped, but more PCBs were turning up than could be accounted for by accidental sources. It was then discovered that, since 1960, Monsanto — the St. Louis company that manufactured all the PCBs used in North America — had been encouraging the use of PCBs in more "open" systems. For example, PCBs increased the effective kill times of certain agricultural pesticides.

Using PCBs in sealants, paints, varnishes, waxes, and even printers' inks made those products more resistant to photochemical break-down. Car tires and brake linings containing PCBs lasted longer. In 1970, when the U.S. Food and Drug Administration (FDA) found high PCB levels in certain breakfast cereals, it was found that the cardboard used to make the cereal boxes had been made from recycled carbon-less copy paper, which contained PCBs. Eventually, the FDA said "Enough," and Monsanto halted its production of open-use PCBs, but by then it was estimated that the company had manufactured more than a quarter-million tonnes of PCBs in the past decade, between 50 and 75 percent of which had "escaped" into the environment.

PCBs were banned in the U.S. and Canada but are still used in Asia. In North America, there are still thousands of organic com-pounds in use whose toxic effects are unknown: a report issued in 1994 by the U.S. National Research Council revealed that there was no information at all on 80 percent of the 50,000 registered industrial chemicals (not including pesticides) then in use. Safe or acceptable exposure levels have been set for fewer than 700 compounds. The rest, in the words of an attorney with the Natural Resources Defense Council, "are assumed innocent until proven guilty." The Environ-mental Protection Agency does not require a manufacturer to prove that a new chemical is safe before putting it on the market: it is up to some environmental group to prove that it has adverse effects.

The studies also showed that PCB concentrations in wildlife tended to be higher the farther north the tests were run — polar bears on Sval-bard and Brougham Island, seals in Norway, peregrine falcons in the Queen Elizabeth Islands, places far from where PCBs were used, showed particularly high levels of contaminants. Organochlorines migrate to the Arctic by a process of evaporation and precipitation similar to the cycle water goes through. Compounds released into the air in temperate or subtropical latitudes move north because they need colder air to keep the cycle going: the aerosols condense into the ocean, where they evaporate into aerosols again, move farther north, precipitate out, evaporate, and so on until eventually they reach water cold enough to retain them as dissolved compounds. On the *Louis*, Eddy described a "thought experiment" in which a glass of warm water containing PCBs is placed beside a glass of cold water containing no PCBs. Before long PCBs from the first glass have migrated into the second; in just such a way volatile compounds migrate from the warm water in the tropics to increasingly colder water to the north. He called the process "a chemical conspiracy" that contaminates an ocean like the Arctic, thousands of kilometres from any economic zone that benefits from their use. Global warming will force these contaminants to move even farther north in search of water cold enough to prevent them from evaporating.

"This northern migration alone makes organochlorines of greatest concern for Canadians," Robie said. "But then you add to that the fact that they bioaccumulate as they move up the food chain, and you have an even more serious problem. Organochlorines in the water get into the phytoplankton, which is eaten by the copepods, which are eaten by the arctic cod, which are eaten by the seals. And each organism keeps all the PCBs it eats: a copepod eats quantities of phytoplankton, a cod eats hundreds of copepods, a seal eats hundreds of cod. Then polar bears eat the seals, the Inuit eat the seals and the bears. Pretty soon you have high levels of PCBs showing up in Inuit mothers' breast milk."

Malcolm and other polar bear biologists have known about pollution in bears for years. Malcolm has been keeping tabs on bears in the Canadian Arctic, and so far those he has sampled near Churchill, Manitoba, have shown relatively low levels of PCBs — about 8 parts per million. Mammals in other northern regions, nearer to eastern Europe and Asia, where PCBs are still manufactured and used, have dangerously high levels: 32 ppm in polar bears from Svalbard, and 70 ppm in seals in the Baltic Sea. If roseate terns are the canaries in the environmental mine shaft, polar bears are the sponges on the chemistry lab floor. "PCB levels in bears," said Malcolm, "can be up to a billion times higher than they are in the water column." All that contamination is contained in the bears' fat and muscle tissue — the parts that Inuit families eat.

Polar bear meat doesn't make up a large part of the Inuit diet any more, but it is still eaten. One evening while Oolatita and I stood on the *Louis*'s deck watching the sun roll along the horizon and talking about Malcolm's polar bear program, Oolat said jokingly: "I wish they'd take it easy on those bears — we eat those things!" He said his people were not likely to stop eating their traditional foods, such as narwhal, seal, walrus, and polar bear, simply because a bunch of scientists tell them that the meat is contaminated.

"They say there are little things in the bears' meat that will hurt us," Oolat said, "but that won't make us stop eating our country food. We need something we can see with our own eyesight."

In *The End of Nature*, Bill McKibben says he can no longer look on nature with a benign and naive eye. Every time he sees a sunset, in the back of his mind is the knowledge that its soft, roseate glow is probably produced by an excess of hydrocarbon particulate matter suspended above the horizon. Likewise, he says, our delight in the patter of light rain on leaves is inflected with the thought of sulfuric acid eating away at the foliage. Sunsets still delight and photosynthesis continues, "but we have ended the thing that has, at least in modern times, defined nature for us — its separation from human society." Now, standing on the deck of an icebreaker with Oolatita in the High Arctic, I felt the deep truth of those words, for we were watching kittiwakes diving for arctic cod.

Our search for contaminants in the Arctic served two purposes. For one thing, we wanted to know what compounds were getting into the northern waters and where they were coming from. Canada and six other circumpolar nations — the United States, Iceland, Greenland, Norway, Russia, and Finland — participate in the Arctic Monitoring and Assessment Program, which has scientists from each country collecting samples to determine all the pesticides, organochlorines, and radionuclides in Arctic waters. The program has been going for several years, but so far only in limited, easily accessible areas. "We have sampled the continental shelves," said Robie, "but until now we have had no data from the Arctic Ocean. And we've no real idea about how contaminants metabolize up and down the food chain. Information from the Seastars as well as from these little arctic cod will give us our first glimpses into what's up here and how it affects the ecosystem."

The second purpose had more to do with global warming than with pollution. Synthetic compounds in the ocean are used to trace horizontal ocean-current movements and vertical convection rates in the water. Because we know exactly when certain chemicals were introduced into the environment — PCBs in 1930, DDT in 1946, even Chernobyl has its own unique radionuclide signature — when their use peaked and when they stopped being used, oceanographers can determine how long water takes to get from where the chemical entered the world ocean to the High Arctic. And they can determine what happened to the surface waters once the chemicals arrived there. If we find the highest concentrations of DDT at 200 metres, say, we know that that water was in the south at the surface sometime between 1946 and 1970 (when DDT was banned for most purposes in North America). That tells us how long it takes southern surface water to reach the Arctic and ventilate to 200 metres.

In most cases, these two roles of chemical contaminants — as circulation clocks and as climate change indicators — go hand in hand. Take CFCs (chlorofluorocarbons, better known as Freons). Fiona McLaughlin and Louise Adamson ran the IOS's Freon-sampling program in the oceanographic lab. They and Mike Hingston and Frank Zemlyak were looking for four different kinds of synthetic organic compounds in sea water: carbon tetrachloride (CCl_4), CFC 11, CFC 12, and CFC 113. Each of these compounds came into the world at a different time. Carbon tet, the universal solvent developed in the late 19th century, is still widely used to reprocess spent nuclear reactor fuel into uranium. Huge amounts of it have been found in aquifers 180 metres beneath the Idaho National Engineering Laboratory, where 61 percent of the United States' transuranic waste is stored in 200-litre drums. CFC 11 and CFC 12 were first produced as refrigerants in the 1940s and were later developed and patented by du Pont as Freon 11 and Freon 12 for use in making foam rubber and rigid foam insulation. The Freon sisters were originally thought to be the perfect industrial compounds, because they behaved themselves under pressure and didn't bond with any other compounds, at least not near the earth's surface, where du Pont executives lived. The company began phasing them out in 1988, after research conducted in 1972 proved that CFCs were responsible for large holes in the earth's ozone layer. They were replaced with Freon 113: with two chlorine atoms instead of one, Freon 113 was thought to be even more chaste than the Freon sisters — and so it is in the lower atmosphere. But the extra chlorine atom makes it twice as promiscuous in the ozone layer.

For decades, CFC production was a multibillion-dollar industry. CFCs' below-zero boiling points meant that in heat exchangers (which is what refrigerators are) they could suck the heat out of a piece of steak in no time flat. They were also useful in making foam rubber because, as virtually inert gases, they didn't react with other chemicals in the rubber or in workers' lungs; all they did was create bubbles in rubber, and then float away into space, which in the 1950s was an infinite repository for things we didn't need any more. The same virtue made them ideal in aerosol spray cans: you could get shaving foam on your face, or ersatz whipped cream on your fake chocolate-flavored pudding, without any disagreeable foaming agents coming out with the goop. But it was their very inertness down here that made them dangerous up there, in the stratosphere, where raw sunlight first hits the earth's atmosphere. Up there, CFCs are ozone killers.

Ozone is a highly unstable form of oxygen. Its name comes from the Greek word *oz*, meaning "smell" (which makes me wonder about the Wizard of Oz). In the troposphere, the lower reaches of the atmosphere 10 to 15 kilometres above the earth's surface, ozone has a refreshing, life-affirming odor. In the stratosphere, 15 to 35 kilometres up, ozone is the gas that makes life on Earth possible. When nitrogen dioxide (NO_2) in the troposphere moves up into the stratosphere, ultraviolet radiation breaks it down into nitrogen monoxide (NO), leaving a single oxygen atom floating around with nothing to do. Along comes an oxygen molecule, O_2, which combines with the free-floating oxygen atom to form ozone, O_3. Fortunately for us, O_3 has the peculiar property of absorbing ultraviolet radiation from the sun. Actually, there are two bands of ultraviolet light: UV-A and UV-B, and the ratio of one to the other is known as the extraterrestrial constant, except that it isn't very constant any more. Ozone blocks all of UV-A and lets some UV-B through. As the ozone layer diminishes, more UV-B gets through. If we are exposed to UV-B for long, we get burned; if we are exposed to too much UV-B, we get cancer. Also, food crops such as soybeans and phytoplankton (which absorb huge amounts of atmospheric carbon dioxide) don't grow, and surface-dwelling zooplankton and fish die. We lose the bottom of the ocean food web.

Until 1974, no one knew that the ozone layer was in trouble. That year, two researchers from the University of California, F. Sherwood Rowland and Mario Molina, theorized that since three loose oxygen atoms seeking to recombine into O_3 constituted such an unstable relationship, it wouldn't take much to convince them to combine into something else. Suppose, they said, some other element were introduced into the equation, an element that combined more easily with oxygen atoms than other oxygen atoms did — let's say chlorine. What would happen? Well, the chlorine atom would combine with one of the oxygen atoms to form chlorine monoxide, leaving the other two oxygen atoms to do what they always wanted to do anyway, which is get back together to form oxygen, O_2. This new O_2 molecule could then be broken up by ultraviolet radiation, creating two new single oxygen atoms to combine with a new chlorine atom to form more chlorine monoxide. Pretty soon we'd have a whole new gaseous composition in the stratosphere, not much of which would be ozone, and a whole lot of UV-B would sneak past it to get to the earth. It was all hypothetical, of course; how could that much chlorine gas get into the stratosphere? Natural chlorine — from the salt in sea-water spray, for example — combined too easily with water molecules to rise farther up in the air than the clouds. We would have to introduce some new form of chlorine that didn't mix with water, or react with any of the hundreds of other elements in the troposphere. As John Firor put it, "If we did wish, for some reason, for chlorine at the earth's surface to move into the stratosphere, we would have to arrange for the emission at the surface of the earth of a chlorine-containing gas. We would, in addition, have to find a chlorine-containing gas that did not react readily with anything, one that was not very soluble, and one that, upon reaching the stratosphere, could be broken down to release free chlorine only by the action of strong ultraviolet light." Such a thing seemed highly unlikely. It didn't exist naturally, except possibly in volcanic emissions, and why would we create such a dangerous and unnecessary threat to our own existence? Cut to Homer Simpson using an aerosol underarm deodorant.

Rowland and Molina calculated that if all the CFCs produced in North America since the first electric refrigerator unit was introduced were to rise up into the stratosphere, a journey they figured would take about 50 years, then 7 to 13 percent of the ozone layer would be destroyed by the year 2000.

The now-famous hole in the ozone layer over the Antarctic was first detected in 1985 by a team of British scientists who had been monitoring global ozone levels since 1957; they reported the anomaly in *Nature* magazine the following year, adding that the layer had been depleting since 1981 but they hadn't wanted to say anything rash until they were sure. They speculated that CFCs might have something to do with the depletion. The presence of the hole was confirmed by a total ozone mapping spectrometer aboard a NASA satellite. After that, things moved rapidly. "Nonessential" — meaning nonindustrial — uses of CFCs were banned. Du Pont voluntarily stopped production of the Freon sisters 11 and 12. And in 1987, representatives of 24 industrialized nations met in Montreal and signed the Montreal Protocol on Substances That Deplete the Ozone Layer, promising to freeze production right away of even essential CFCs at 1986 levels, and to reduce that by 50 percent by 1998. The initiative had some salutary effects. One of the biggest users of CFCs in the United States was the electronics industry: aerosol spray cans were great for cleaning delicate or hard-to-get-at equipment. When the U.S. government ordered the industry to reduce its use of CFCs as cleaners, IBM came up with a startling new substitute: soap and water.

Still, since most of the 20 million tonnes of CFCs are still in the atmosphere, and will be for the next 50 years, the depletion of the ozone layer is expected to continue unabated until 2030 even if everyone suddenly stops producing them, which of course everyone hasn't. The Montreal Protocol allowed nonindustrialized nations that didn't have their fair share of CFCs to go on producing them (or importing them) until they'd caught up with the 1986 levels of the industrialized nations. And as long as the ozone hole was confined to the Antarctic, people living in the Northern Hemisphere continued feeling okay about installing air conditioners and shipping crates of hair spray to Indonesia.

Then, in February 1989, a second hole in the ozone layer was discovered, this one over the North Pole. At a hurriedly called meeting in Helsinki, 80 countries signed a second accord agreeing to phase out CFC production altogether by the end of the century. But ozone levels in the northern latitudes continued to decline, and at a rate much faster than had been predicted. Stratospheric ozone levels over Canada in 1993 averaged 14 percent below normal in the spring and 7 percent below normal for the summer: the level that Rowland and Molina had predicted, but seven years ahead of schedule. The explanation was thought to be volcanic chlorine from Mount Pinatubo, but if that were the case, as the authors of the most recent *State of the Environment Report for Canada* observe, "the reason for similar large depletions over southern Canada in the spring of 1995 . . . is unclear." The reason is not unclear to everyone. As environmental researchers point out, we are not even close to meeting the terms of the Montreal Protocol, let alone those of the more stringent Helsinki Accord. Levels of carbon tetrachloride in the atmosphere, for example, are still increasing at the rate of 1 percent a year.

It is true, however, that the production of CFCs 11 and 12 has been halted, and that is good not only for the environment but also for geochemical oceanographers like Fiona and Mike. "We're accustomed to thinking of Freon as a nasty gas," said Mike, "and it is when it's in the air. But in sea water it's a very useful tool for determining ocean circulation," and thus for calculating the rate of global warming.

Fiona, Louise, Mike, and Frank operated nearly identical machines in the wet lab, large stainless steel boxes connected by several kilometres of tubing. Their banks of equipment occupied most of the lab's counter space. The idea was to add a thin stream of nitrogen to sea water taken from various depths of the ocean. "The CFC molecules suspended in the water see the nitrogen stream and say, 'Hey, here's a way out of here, let's hitch a ride,'" Fiona explained. When the CFCs hop on the nitrogen train, they are whisked into a dryer, or desiccator, where the water is evaporated away, leaving only the CFCs and nitrogen. The gases are then concentrated and directed into a trap, which compresses them into an even-flowing stream, which is then pumped into a gas chromatograph.

"It's a sort of glorified superefficient distillation plate," said Mike, as if that explained everything. "The gas chromatograph contains 75 metres of glass tubing coated on the inside with a material that picks up CFC molecules as they separate out from the nitrogen mix. As the mix passes through the tube it is heated. Since they all have different boiling temperatures — Freon 11 boils at minus 4 degrees Celsius, carbon tet at plus 80, and so on — they all boil off at different times."

All the computer had to do was record how much CFC came out at each stage of the heating process, and the scientists could figure out how much of each there was. "Up here," said Fiona, "there isn't a lot. We have to pump hundreds of litres of sea water through the thing just to get a minimal reading, and this equipment is so supersensitive that if someone down in the engine room uses a can of WD-40, we'll get a Freon spike showing up in the data. Even at that, we're looking at a few picamoles per litre."

On this trip, Fiona and Mike began picking up high concentrations of CFC-laden Atlantic water almost as soon as we left the continental shelf and began descending to the Chukchi Abyssal Plain. By station 10, at 76 degrees, Fiona found that the CFC tracers showed more ventilation and deeper penetration than she had found on the *Larsen* in 1993, indicating that the Arctic Ocean was ventilating very rapidly in that region. This was consistent with the temperature evidence Eddy and Jim Swift were turning up. As we moved farther out over the Makarov Basin, the warm water was moving with us. There was no front — the abnormally warm Atlantic Layer was now spreading throughout the Canada Basin, or so it seemed. How much warmer it would get, and how far into the Canada Basin it had spread, remained to be seen. But it was already safe to say, as Rachel Carson had warned in 1950, that the top of the world was warming up. And not only because of an increase in greenhouse gases in the atmosphere. This was an entirely new phenomenon, one that nobody had foreseen. The Arctic Ocean was warming up from below.

12

Dirty Ice

But not yet have we solved the incantation of this whiteness, and learned why it appeals with such power to the soul. . . . Is it that by its indefiniteness it shadows forth the heartless voids and immensities of the universe?

– HERMAN MELVILLE, *Moby-Dick*

WE CROSSED 80 degrees N during the night of August 5, at 173 degrees w longitude. It was a landmark, or rather an icemark: on most maps of the Arctic Ocean, which show the North Pole at the center, the lines of longitude radiate in from the edges and stop at a circle formed by the 80th degree of latitude. Within that circle there is nothing but a central dot, as though that's where the mapmaker put his compass point. I'm sure this is primarily for the clarity of the map — if the lines continued and converged at the Pole, the result would be a big dark blob, very messy. But I also suspect that the mapmaker assumes that no one really needs the lines up there, that anyone who goes that far north is beyond the need of geography, that they are lost anyway. The effect on the mind is of abandonment, that all things connected to the real world, mundane things like lines of longitude, have come to an end. Follow the 173rd meridian south from the 80th parallel and (after breaking through the Aleutians) you would hit Hawaii (almost), Western Samoa, part of the Cook Islands, and the southern tip of New Zealand; follow it north and you enter a vast, howling emptiness, a magic circle, a blank, a void. And at the center of the white void, the hole made by some anonymous cartographer's compass: the North Pole.

We had had tough going during the night. I stayed on the bridge until 2 A.M. with Eddy and the captain as the ship made run after run at ice that kept pushing us back into our own wake, as though the 80th parallel were a physical as well as a psychological barrier. Knut was aboard the *Polar Sea*; as expedition leader, he was trying to divide his time evenly between the two ships. The American icebreaker waited a cable-length behind us, invisible in the fog, as we bashed headlong at the ridge. Whether we would continue on to station 19 and attempt Art's seismic tow depended to a large extent on how successfully we negotiated the ice here, since this was what we would come back through.

At breakfast, I sat beside Caren. She had been up since 6, studying a new TeraScan map. I asked her how the 500-square-kilometre floe of multi-year ice had behaved overnight.

"The Mother Floe," she said. "It split in half, like the Red Sea." She pushed the map across the table for me to look. "Right down the middle, and right along our path to station 19."

It was true. The large, angry red blob that had lain directly between us and the Canada Basin was now two smaller red blobs. It was like looking at a photograph of cell division, or blood-platelet separation. Red Sea, indeed: though now composed of somewhat smaller floes, the color-coded ocean ahead of us was still 100 percent red, covered by multi-year ice. Except for the open path between the split floe, which might be good news for Art, there was still a lot of heavy ice to get through.

The fog lifted at 10 o'clock, and we stopped for station 18, but we were not where we had wanted to stop. Captain Grandy ordered an ice recon. Knut and Art were flown over from the *Polar Sea*, and they and Fred, with Caren's map on his clipboard, made an ice reconnaissance flight in a line directly ahead of the ships, to the northeast, while the rest of the scientists conducted a hastily contrived station at 80 degrees N. We were 2,800 metres above the eastern foothills of the Mendeleyev Ridge, so close to the Canada Basin that we could sense its presence below us. If the Mendeleyev Ridge were the Rockies, and the Canada Basin were the prairies, then we were in a holding pattern somewhere over Calgary.

While the helicopter was gone, the crew lowered the rosette from the starboard boat deck. The winch unwound at the rate of 1 metre per second, and it took 46 minutes for the rosette to reach the bottom. In the milking shed I stood with Jim Swift and watched the samplers filling their sterilized glass bottles. Jim said he was looking forward to getting to station 19 as eagerly as Art was.

The extent of the warm-water anomaly was still unknown, and although the upper ocean layers in this part of the Canada Basin had been sampled before (from Soviet ice stations), Jim was eager to find out if the temperature here had risen as well. "Besides," he said, "no one has ever sampled the bottom waters. You have to remember that the Canada Basin is the only significant ocean basin in the world that has never been transected. It makes excellent scientific sense to go in there, and I hope we can make it."

Jim wasn't totally convinced that the measurements from the *Larsen* trip indicated a large-scale incursion of warm water spreading through the western half of the Arctic Ocean. "There's a significant chance that they just happened across an isolated bolus of water." A bolus, or eddy, is a relatively small, flat pancake of warm water that spins across vast expanses of ocean at various depths, much as a Frisbee spins through air. When the Gulf Stream makes its sharp right turn up the Atlantic coast off Nova Scotia and heads east toward England, it sends dozens of these eddies swirling out over the Grand Banks. Jim thought something like that might account for the warm temperatures the IOS team had found in the Canada Basin. "The Arctic basins are well known for this," he said. He added that he was "duly impressed" with the warm water in the Chukchi Borderland, but was reserving judgment.

We both looked out through the shed door at the unforgiving expanse of frozen ocean. It was no longer a flat prairie of ice; pressure ridges had thrown up huge chunks everywhere. For a seismic tow, each ridge would have to be traversed in a perfectly straight line at a constant speed by the *Louis*, while the *Polar Sea* followed closely behind, towing a kilometre-long cable attached to an array of delicate seismic instruments.

"It'll be like trolling in a swamp," I said to Jim.

"We have to accept that we can really only go where the ice is not."

The helicopter returned at 11:30, and Captain Grandy, Knut, Art, Eddy, and Caren got together in the captain's dining room. I sat with Fred and Steve Hemphill next door in the officers' mess, drinking coffee and sterilized milk and watching the ships' stewards setting up the tables for lunch.

"Well?" I asked Fred. "Rumor has it we're not going to station 19. Too much ice."

There had been whisperings among the scientists that morning — Captain Grandy was reportedly "leaning against" going to station 19; a new satellite map had come in showing ten-tenths coverage straight ahead and only eight-tenths to our west, back up the Mendeleyev Ridge. And Mike Hemeon, the first officer, had hinted that there might be new plans posted. "Even if we could get to station 19," he had said, "doing a seismic tow in heavy ice is a problem. There is a great risk of damaging the ship."

Fred said the ice ahead of us was much like the ice we'd been sailing through for the past two days. "More pressure ridges, maybe, but also some patches of open water." The Red Sea floe had indeed parted up the middle, but there was a lot of pressure in the ice, and the floe could easily close up around us when we were halfway through it.

"We might go in thinking we're the Children of Israel," said Steve, "only to find we're really Pharaoh's army."

On the other hand, the open water suggested that the pressure might be easing up. "So it's one of those 50-50 decisions," said Fred, shrugging. "We'll just have to wait and see what the captain decides to do."

We found out after lunch at the science meeting. We'd skip station 19. The *Louis* was going to stay here at station 18 overnight to do maintenance on her engines while the *Polar Sea* did a series of piston cores going down the slope of the Mendeleyev Ridge. The cores were a consolation prize for Art; they would be the most northerly cores ever made in the Canada Basin, but would tell him nothing about the whereabouts of Charlie. The two ships would rendezvous at 8 o'clock the next morning, then head back south together to pick up the Seastars.

"From there," said Eddy, "we'll maneuver west along our present line of latitude to do a short station at 176 degrees longitude."

"How short?" asked Ron Perkin. Ron had been doing a series of mini rosette casts for the past two weeks, taking a helicopter and a small CTD a kilometre or so out from the ship's track line to get a wider swath of data, and he needed to know how long the ships would be stopped so he could plan how many casts he could get in.

"Two to three hours," Knut answered.

"Staying in this hard Canadian ice," said Mike Hemeon.

"After that," Eddy said, "we'll cross the international date line and get into some of that first-year ice coming up from the Siberian shelf."

Station 18 was completed at 8 o'clock, and the *Louis* wasn't going anywhere until morning. The bar, which was open nightly from 8 to 10, was full of thoroughly chilled scientists and ship's officers trying to warm up. Captain Grandy had called for a "Half-Birthday Party," since, geographically, 80 degrees was halfway to the Pole. The galley sent up half a birthday cake with "Happy Half-Birthday, Eddy and Mary" written in white on the chocolate icing.

"What's a half-birthday?" I asked Eddy.

"Well, my birthday was really six months ago," he said.

I asked Mary Williams the same question.

"Well," she said, "I only half admitted it was my birthday."

Ron Ritch was standing by the bar. He was glad the ship had stopped for a while, he said, because whenever we moved he always found himself listening for data points.

"We thought of calling them something more interesting," he said, "like 'impact events,' or 'collisions with ice,' but we decided that was too dramatic. We didn't want to alarm anyone. So we settled on 'data points.' Nice and technical. No need to worry, Captain, that was just a data point."

"How many data points have you recorded so far?" I asked.

"Four thousand."

"Four thousand data points? We've hit four thousand chunks of ice?"

"Oh, far more than that. We only record the major impacts, those over a certain threshold of force. We measure them in tons per square inch. We've recorded impacts up to about 1,500."

"How would you describe an impact of 1,500 tons per square inch?"

He thought for a while. "I'm not very good at this kind of thing," he said, "but roughly speaking, it would be like dropping four fully loaded Mack trucks onto a kitchen table."

"That would account for the noise in my cabin," said Darren. He and Dave shared a cabin on the same deck as ours. "It's like trying to sleep in a cement mixer. And then there's that bubbler compressor. It makes my ears bleed."

Frank Zemlyak agreed that pumps could be extraordinarily noisy. "There was one on the *Oden*," he said, "that was so big it could fill a heeling tank in seven seconds. I did some rough calculating one day. I figured that if there were 186,000 bathtubs in Halifax, the *Oden's* heeling pump could fill them all in ten minutes."

"Had a lot of time on your hands during that cruise, did you, Frank?" said Darren.

Frank laughed. He said he had also developed a program for his computer that would give the distance from the North Pole to any co-ordinate on Earth. "We can put up one of those tourist signs at the Pole — get an arrow and paint 'Halifax, 4,324 kilometres' or whatever on it."

"Is that what you did on the *Oden*?"

"Everyone does it," said Frank. "The Pole would be littered with signs if they didn't drift away with the ice. It would be like the top of Mount Everest. Everyone can put up a sign with his or her hometown on it, and the distance."

Everyone told Frank what a tacky idea that was and made a mental note to give him the coordinates of their hometowns.

The next day, August 7, Stefan and I helicoptered back to the *Polar Sea*. The mood on the American ship had deteriorated since the cancellation of Art's seismic tow. We arrived at lunchtime and found most of the scientists in the crew's mess, eating grilled CheezWhiz sandwiches off aluminum trays and watching *The Exorcist* on television. After lunch, at a meeting in the chartroom with Terry Tucker and Pat Wheeler, Art recounted some of the deliberations that had taken place on the *Louis*. There was some concern about the short time the scientists on the *Sea* had to prepare for upcoming stations.

Terry, who works for the U.S. Army's Cold Regions Research and Engineering Lab (CRREL) in Hanover, New Hampshire, was coordinating the ice program on the American ship. He said he and his on-ice team would probably have to take a pass on the next station. "There isn't anything much we can do in two or three hours." One of the team members, Erk (pronounced Erik) Reimnetz, would make a brief survey of the ice, and if he found anything interesting then Terry would request more time.

Erk was a tall, blond, German-born scientist who worked for the U.S. Geological Survey. He went out onto the ice whenever he could to collect jars of dirt that he scraped off the ice surface. He had been doing it for more than 25 years; he'd been one of the first scientists anywhere to realize that grains of sand and sediment on top of ice floes reveal a great deal about ice formation and movement. Anchor ice, which forms close to the coast, in shallow water over the continental shelf, grows upward from the ocean floor and brings up sediment as it rises to the surface — sometimes as much as 1,000 tonnes of sediment per square kilometre of ice. When anchor ice travels out into the Arctic Ocean to become sea ice, the sediment travels with it. Sediment from each section along the Siberian coast has its own signature of minerals, chemicals, and, in some cases, radionuclides. By sampling dirty ice in the middle of the ice pack, Erk could tell where along the long Arctic coastline a particular floe had originated.

He showed me a map of the Arctic Ocean he had made, with hundreds of arrows depicting ice-flow patterns. The ice seemed to swirl in two huge circles, with some interesting exceptions. Ice with Siberian dirt and even trunks and branches of Siberian trees routinely shows up along the Alaskan and Canadian coasts, indicating that some ice from Russia moves eastward, hugging the shore; but dirt and wood from other parts of Siberia have been found in ice coming out of the

Arctic through Fram Strait — just as bits of the *Jeannette* had in 1892 — which meant that some Russian ice veers away from the coastal currents and drifts northward, across the Arctic Ocean. Erk wanted to know which ice was likely to go where.

As luck would have it, the *Polar Sea* had stopped in a vast field of dirty ice. Erk and Terry both got off to take samples, and the rest of us decided we might as well go too. So far, walking on ice had been like walking across a snowy landscape, with an occasional climb if the terrain were hummocky. Arctic snow is more like crushed glass than the fluffy stuff we know farther south, but at least it is bright and shiny. At this station, it was as though someone had scraped all the snow off a dirt road, exposing the sandy ice and water-filled potholes below. It looked like a Toronto parking lot in March. It wasn't easy to walk on, because it was dotted with dozens of melt ponds — pools of clear, fresh water, melted by the wan Arctic sun, resting in grit-filled depressions in the dirty ice.

Terry carried a small, hand-held radiometer attached to a 3-metre boom, which he stuck out over several of the ponds as part of Caren's experiment to determine the difference in brightness temperatures of melt ponds and ice. Terry wanted to know how melt ponds contributed to the overall Arctic heat budget. Bright, shiny surfaces reflect more solar radiation than they absorb; dark, porous surfaces absorb more than they reflect. In the south, snow reflects more energy than plowed land, which is partly why we have long winters on the prairies. Snow conspires in its own continuance. In the tropics, where we have cut down vast areas of bright, shiny leaves and replaced them with dark, absorptive cropland, we have drastically affected the earth's heat budget: as the dark earth absorbs more heat from the sun, it contributes to global warming.

In the Arctic, ice reflects more radiant energy than open water does, and so diminishing the extent of the ice cover (by global warming) actually encourages more global warming. Terry wanted to know whether the ice pack absorbs more solar radiation in summer, when its surface is dotted with melt ponds, than in winter, when the pack is a uniform white. Logically, it should — water absorbs more radiation than ice — but logic without data is just hunch. A related question is whether the color of the ice, as seen through the melt ponds, makes a difference in the reflectivity of the melt pond.

"Some ice ponds are blue," he said, "and some are green, and I can't figure out why. I tried to determine whether it has anything to do with how fresh or salty the water is in the ponds, whether it's a result of a change in the ponds' albedo." Albedo is the ratio of the amount of radiation reflected by a surface to the amount it absorbs. Melt ponds do not have a constant albedo: as their composition changes, so do their reflective properties. Ponds are usually fresh, but in late summer, the bottom of a pond can melt so far into the ice that it actually breaks through the bottom of the floe, creating a small hole like the drain in a sink. When that happens, salt water and sometimes arctic cod seep up into the pond. "I wondered if the different color was a result of the melt ponds' retaining more heat because of their salt content. But I've measured blue melt ponds with lots of salt in them and blue melt ponds with no salt in them, and green melt ponds with a layer of salty water sandwiched between two layers of fresh water, and I've found no difference in their albedos. So I don't know what to think."

Not having a theory to test didn't stop Terry from continuing to collect data. "I guess I'll just keep measuring away until another brilliant idea occurs to me." Besides, he still had to compare melt-pond albedo with that of snow-covered winter ice. If melt ponds absorbed more heat from solar radiation than a frozen ice surface, they, too, could act as positive feedbacks for global warming.

We found Erk squatting on the ice a few hundred metres from the ship, scraping dirt into a small glass jar with a spoon from the *Polar Sea*'s dining room. The hillock of ice he was sampling was completely covered by sediment. As the sun melts the ice at the top of a floe, bits of grit and sand embedded in it become concentrated on the surface. Walking over such a crust can be tricky; there is usually ice beneath it, but it can be a long way down. Erk has taken thousands of sediment samples from the tops of ice floes and compared them with samples from the ocean floor all along the Siberian coast. He could tell if a floe sampled here originated in the Laptev Sea or the Kara Sea, or whether it contained sand from the Lena or the Yenisei River.

"There is a huge polynya in the Laptev Sea," he said. A polynya is an area of water in the Arctic that is never covered with ice; in the larger polynyas, new ice continuously forms but is immediately shipped out to the edges. Polynyas are all over the Arctic. The biggest one, the famous North Water at the top of Davis Strait, was discovered in the early 19th century and kindled the Open Polar Sea debate. The *Polar Sea*'s previous expedition had been to the Greenland Sea polynya, where Lisa Clough had pulled up creatures from the sea bottom she had never seen before. The Laptev Sea polynya is huge, nearly 18,000 square kilometres, and very shallow. "It is a veritable ice factory," said Erk. "It makes ice at a fantastic rate. Normally, ice grows to about 2 metres in thickness in its first year: the Laptev Sea polynya can create ice 2 metres thick in one day. And because it is so shallow, new ice can be incorporating sedimentary particles and transporting them into the High Arctic all winter long."

Erk's years of sampling indicated that ice from the Laptev Sea normally entered the transpolar drift, the stream of arrows on his map that went from the Siberian coast across the North Pole and out Fram Strait — Nansen's ship was probably stuck in ice that had been born in the Laptev Sea polynya. But since 1988, more and more Laptev Sea ice had been turning up in the Beaufort Sea. "I don't know why," he said. "But it is important to know where that ice goes, because some of it contains trace levels of radionuclides, especially plutonium. It's difficult to test for plutonium, because I need 10 grams of sediment for the test, and normally it's very difficult to get that much from a single floe. But this is fantastic," he said, indicating the piles of mud around him. "This is very fortunate. It is easy to differentiate between radionuclides produced by bomb testing and radionuclides from nuclear reactors. When I get these samples analyzed, I'll know exactly where all this beautiful dirty ice is coming from."

The Soviet Union conducted extensive nuclear bomb tests on the two islands of Novaya Zemlya, in the Kara Sea, mostly between 1955 and 1962 (although the tests continued until the collapse of the Communist regime in 1990), and fallout from them is still polluting Kara Sea water. A recent report by Vitaly Adushkin, of the Russian Academy of Sciences, and Gennady Krasilov, of the Russian Committee of Hydrometeorology, based on information that has only been available since the collapse, documents a total of 132 tests, 87 of them prior to 1962. According to Adushkin and Krasilov, "the total yield [in terms of radioactive pollution] of air and undersea nuclear tests at the Novaya Zemlya Test Site is approximately twice more than the total yield of the U.S.A., England, France . . . and China." These explosions have created high levels of three distinct radionuclides: cesium 137, strontium 90, and cobalt 60 — the signatures of bomb testing.

Of equal concern is evidence presented by Aleksei Yablokov, Boris Yeltsin's senior environmental adviser, that the Soviets regarded the water surrounding Novaya Zemlya as a great place to get rid of unwanted nuclear reactors: at least 15 nuclear submarines and one icebreaker, the *Lenin*, the U.S.S.R.'s first nuclear icebreaker, have been scuttled in the Kara Sea, their reactors filled with concrete but still intact, six of them still fully loaded with radioactive fuel. Two more submarines were sunk in the Sea of Japan. In all, 2.5 million curies of radioactive waste have been dumped into Arctic waters, more than twice as many as have been created by the next 12 nations on the world's list of nuclear polluters.

Of considerably greater concern are the huge quantities of radioactive wastes entering Arctic waters from nuclear reactors in the Siberian interior. According to *The State of Habitat Protection in the Arctic*, a report published by Norway's Directorate for Nature Management in 1994 (which I borrowed from Captain Brigham's well-stocked library), the Arctic — "the world's largest remaining wilderness" — represents a total area of 13.6 million square kilometres. Canada has the biggest share of that, with 5.3 million square kilometres, or 49.2 percent of Canada's total landmass. Russia is second, with 4.8 million. Very little of that is protected — 6.2 percent of Canada's Arctic, 3.4 percent of Alaska's, and virtually none of Russia's. The various new Russian governments are busy trying to catch up — according to one estimate, there are currently 80,000 environmental assessments under way in the former Soviet Union — and certainly no country in the West can cast the first stone at Russia. But it might well be too late for any reversal in the Russian Arctic.

The Norwegian report says that nuclear waste is the major threat to the Arctic ecosystem, more serious to the future of Arctic wildlife than contamination from pesticides and industrial chemicals. And the biggest sources of radionuclides entering the Arctic are the Siberian rivers emptying into the Kara and Laptev Seas.

One of the most disturbing matters to come to light has been the appalling record in the handling and storing of the former Soviet Union's nuclear waste. According to D. J. Peterson, whose book *Troubled Lands: The Legacy of Soviet Environmental Destruction* was published in 1993 (another loan from Captain Brigham), the Soviet government had pretty well written Siberia off as an irredeemable wasteland, fit only for mineral exploitation, dissident labor camps, and toxic waste sites. The polluting of Lake Baikal, the largest freshwater lake in the world (containing 25 percent of the world's fresh water), is well documented.

But less known is that deep in the Siberian interior there were at least three "secret cities" where nuclear material for the U.S.S.R.'s bomb testing program was produced. The cities had no names and were known by their postal codes: Tomsk-7, for example, was located in a postal district 24 kilometres north of the provincial capital city of Tomsk, on the river Tom (a tributary of the Ob, which flows 2,500 kilometres north to empty into the Arctic Ocean). Tomsk-7 was built in 1958 as the U.S.S.R.'s main nuclear processing facility; because no one outside and hardly anyone inside the Soviet Union knew of its existence, Tomsk-7 was accountable to nobody. Greenpeace never made it to Tomsk-7. Russia's own environmental investigators now estimate that between 1958 and 1990, the Siberian Atomic Energy Station at Tomsk-7 dumped 115,000 tonnes of solid nuclear waste and 33 million cubic metres of liquid radioactive waste directly into the Tom River, where it flowed into the Ob and, eventually, into the Kara Sea. The Moscow newspaper *Izvestiya* reports that 38 people in the Tomsk area died of radioactive poisoning in 1990 after eating wild game they had caught near Tomsk-7.

Another secret city, Chelyabinsk-40, on the Techa River, also a tributary of the Ob, has an even worse record. The Mayak nuclear facility in Chelyabinsk-40 was the birthplace of the Soviet nuclear weapons program. Its two reactors, known cryptically as Object A and Object B, produced the U.S.S.R.'s first atomic bomb, which was detonated on Novaya Zemlya in August 1949. By the mid-1950s, Mayak had dumped millions of curies of radioactive waste directly into the Techa. As early as 1955 the Soviet government ordered the dumping stopped, had the entire length of the Techa River fenced off, and evacuated 7,500 people from villages along its banks.

Mayak continued to produce nuclear weapons, however, and stored its waste in nearby Lake Karachai instead of dumping it into the river. Within two years it had dumped 120 million curies into the lake (by way of comparison, the meltdown at Chernobyl produced 50 million curies). By 1957, Lake Karachai was so contaminated that merely standing beside it for an hour was enough to give a person a lethal dose of radiation. So that year a third disposal method was adopted: Mayak built 16 concrete storage tanks and buried them 30 metres underground. The tanks were kept from overheating by an elaborate cooling system that broke down almost immediately: on September 29, 1957, one tank heated to the point of uncontrolled nuclear fission, and the resulting explosion spewed 2.1 million curies of cesium, ruthenium, and strontium more than half a mile into the air, spreading it over an area inhabited by 270,000 people. Only about 10,000 of them were evacuated. The villages they lived in were bulldozed and their names removed from the map.

Ten years later, a drought in the Ural Mountains caused the water level in Lake Karachai to drop significantly. This exposed a metre of radioactive waste on its shores, which dried and turned to dust. Wind storms blew airborne strontium and cesium over a 2,000-square-kilometre area: 935 people died from radioactive sickness, and 437,000 were exposed to levels high enough to cause cancer. In 1990, *Moscow News* reported that "not a single large city was evacuated, because it was too expensive to resettle large numbers of people." Even now, the paper lamented, "people continue to grow contaminated food which they eat themselves and ship to the cities."

The earliest warning that the Soviet nuclear program was in trouble came in 1965, when the U.S. Coast Guard cutter *Northwind* conducted the first scientific expedition into Siberian waters. Despite the severe and oft-expressed disapproval of the Soviet Union, the *North-wind* sailed from Copenhagen on July 15, rounded Norway, and chugged into the Arctic, keeping well outside the Soviet Union's 12-mile limit. Civilian scientists aboard conducted oceanographic and water chemistry sampling in the Barents and Kara Seas, "for which," stated the Coast Guard press release, "scientific data is virtually nonexistent in the U.S." The results confirmed that high levels of radioactive waste were reaching the Arctic Ocean. The levels were so high, in fact, that military analysts in the United States were able to determine the kind of nuclear weapons that would create such waste — which of course was the purpose of the expedition. Subsequent monitoring of the area has confirmed that the Kara Sea and the water south of Novaya Zemlya are the most radioactively contaminated spot on Earth.

On the *Louis*, Kathy Ellis and Rick Nelson, both of the Bedford Institute, and Brenda Ekwurzel were conducting the radionuclide programs. They were actually studying two kinds of radioactivity: natural radioactivity and artificial radioactivity. It is the nature of radioactive material to break down, and it always breaks down in the same way. Two natural sequences of radioactivity occur in sea water: radium to lead 21 to plutonium 210; and uranium to thorium 234. The product of a radionuclide's decay is called its daughter. Radionuclides and their daughters are found primarily in the near-surface waters along the continental shelf, adhering to things suspended in the water and eventually settling to the bottom. Since the rate at which each radionuclide becomes its own daughter is constant (the material's half-life), the nuclear family is useful in giving us an idea of the rates of flux in the water — the rate at which particulates such as carbon settle to the bottom. "The rate of thorium removal from surface water is related to the prime productivity rate," said Kathy. "If lots of thorium is being removed, that means there are lots of particles in the water removing them, and that means there is a lot of production taking place." In other words, lots of thorium in sediment samples means there must be lots of phytoplankton in the surface water.

The artificial, or anthropogenic, radionuclides come from bomb testing and nuclear reactor sites. The radionuclides involved — some of them from sites outside the former Soviet Union, such as England and France — are cesium 134 and 137, strontium 90, plutonium 238, 239, and 240, and iodine. Find one of these isotopes in Arctic sea water, or rather a ratio of one daughter isotope to its parent, and Kathy will be able to tell you more or less where the water came from and how long it took to get there. That is the beauty of isotopes: they decay at specific and known rates. Cesium 137 takes exactly 30 years to decay into cesium 135, so if you find an amount of cesium and determine that half of it is cesium 137 and half is cesium 135, then you know that that cesium was released exactly 15 years ago. And each release site in the world has its own telltale radionuclide. Changing world climate shows up as changes in the makeup of Arctic sea water; so do changing world politics. "Before Chernobyl," she said, "almost any time you found cesium 134 in the water, you knew it came from the Sellafields reactor in England, which until 1975 dumped most of its nuclear waste into the Irish Sea. But Chernobyl has put a lot of new cesium 134 into the atmosphere, especially in the North, so now you can't be sure any more." Kathy was on an ice island in the Canadian Arctic, measuring cesium 134 in the atmosphere, when Chernobyl went up. Her instruments jumped like scared rabbits as the radioactive cloud passed overhead. "My data points took a terrific spike upward," she said. "I could practically chart the movement of the front."

The artificial radionuclide program sought to determine what radionuclides were there and where they probably came from: Atlantic water (i.e., from France and England, up the Gulf Stream, and through the Barents Sea) or Russian water (seeping up from the Kara and Laptev Seas). This would give the oceanographers time sequences for water movement along those highways. For instance, a certain ratio of cesium 134 to cesium 137 in the Makarov Basin would indicate that its water came from the Kara Sea and had taken 30 years to get there. That is useful information when trying to calculate how long the effects of global warming will take to get into the High Arctic, or how long decreased salinity in the Arctic Ocean will take to be felt in the North Atlantic. The depth at which the isotopes were found was also important: Brenda uses the ratio of oxygen 18 to oxygen 16 to distinguish river water from melted ice water. Water coming from rivers that debouch into the Arctic have a characteristic oxygen ratio, "and if that ratio is found in the surface water, for example," she said, "then we know that it came from river inputs," as would other, more harmful isotopes found in the same samples.

Brenda was also looking for traces of tritium (a daughter of hydrogen), helium 3 (a daughter of tritium), and carbon 14. Like cesium, tritium came into the environment from nuclear weapons tests in the 1950s and '60s, and is now found throughout the Arctic, albeit in small doses ("I'm taking half the Arctic Ocean home with me," she said). After only 12.5 years, it turns into helium 3, so that determining the ratio of tritium to helium gives residency time. Paleontologists use isotopes such as carbon 14 to date fossils; oceanographers use them to date water. On an earlier trip to the Canada Basin, Eddy and Robie had used carbon 14 to learn that the deep water in the Canada Basin has been down there for 400 years.

What Eddy wanted to know next — what he wanted Brenda with her radionuclides, or Fiona with her CFCs, to tell him — was how long the middle of the Atlantic Layer had been where it was and whether it was staying put. He knew that the front was spreading out over the Canada Basin, maybe out over the entire Arctic Ocean. He needed to know whether it was staying down at the 400-metre level, or rising up to the surface. If the latter, it would increase the temperature of the Arctic Mixed Layer enough to lessen the amount of ice that formed on it.

13

Losing Contact

"We're going to discover the North Pole."

"Oh!" said Pooh. "What is the North Pole?" he asked.

"It's just a thing you discover," said Christopher Robin carelessly, not being quite sure himself.

— A.A. MILNE, *Winnie-the-Pooh*

HIGH IN THE CORNER of the crew's mess on the *Polar Sea*, an old John Wayne movie was playing on the TV monitor. Wayne, his trademark oversized bandanna tied around his neck like a bib, had grabbed a shady-looking lawyer by his string tie and was slapping him in the face.

Wayne: "Whaddaya done with my deed?"
Lawyer: "You don't think I'd be stupid enough to keep it here, do you?"
Wayne (slapping him again): "Where is it?"
Lawyer: "Over there in the cupboard."

Art Grantz sighed. He was eating something that looked like hamburger patty and mashed potatoes. It was late at night, long after midnight, and the only food available from the mess was "mid-rats," short for midnight rations (although more colorful derivations had been proposed). The chairs in the mess were attached to the tables, like school desks, so that when the torpedo struck, chairs and tables and hamburger patties wouldn't go flying off in different directions. We had been joined by Erk and Terry, and the three geologists were discussing the Alpha Ridge, a northerly extension of the Mendeleyev Ridge that eventually rose up out of the ocean and became the Queen Elizabeth Islands. We would be passing over it in a day or so. Art was expanding on his theory that the Queen Elizabeths had once been in the East Siberian Sea.

"The Atlantic Mid-Ocean Ridge extended into the Arctic Ocean about 55 million years ago," he was saying, "and it forced the Lomonosov Ridge away from Siberia, pushing it out into the middle of the Arctic basin. A few Russian and American geologists have studied it, but the basin on this side of the Lomonosov is almost unknown. Oh, we know a bit about the bathymetry, we can map the ocean-floor features, but we don't know anything about what caused them, or what's under the sediments."

The Russians would fly their scientists up to an ice station and leave them there, sometimes for years, to drift with the ice until they eventually emerged in the Greenland and Barents Seas. This meant that the scientists didn't have a lot of control over where they did their sampling. They simply set up their tents and equipment and took measurements and soundings as the floes carried them southward to dissolution. Art had a warm admiration for the Soviet scientists. "Icebreakers would be sent up to get them in Fram Strait just before the ice melted out from under them." It was one of these icebreakers, the *Arktica*, that had, in 1977, become the first surface vessel to reach the North Pole. "The *Arktica* went up to rescue a group that had got in trouble at around 88 degrees north," Art said, "and then her captain said, What the hell, we're this close, we might as well go to the Pole. So they did."

"Taking the eastern route," I said.

"Yeah, but still," said Art, "it was a hell of a feat. Anyway, we have fossils from Siberia and the Queen Elizabeths that match, we have geology that matches, and in 1988 we got piston cores with bedrock in them that we could date, tying Tuktoyaktuk in with the Northwind Ridge, and all of that indicates that the Canada Basin was formed by sea-floor spreading in the late Lower Cretaceous, about 100 to 120 million years ago."

There was an explosion of gunfire above us as John Wayne calmly lifted his head above a wagon wheel and, wincing, let off a round of pistol shots.

"But if that's the case," Art continued, "if the Asian continent was once attached to the North American continent, and then spread apart to create the Canadian Basin, you'd think the Canadian Basin would be fairly flat and featureless. But it isn't. It has all these ridges and troughs in it, like the Alpha Ridge. What's it doing there?"

I had talked about this aspect of the Alpha Ridge on the *Louis*, with Peter Jones. In 1983, Peter had participated in the Canadian Experiment to Study the Alpha Ridge (CESAR). "We wanted to call it the Canadian Arctic Experiment to Study the Alpha Ridge," Peter said, "to get the spelling of CAESAR right, you see, but cooler heads prevailed." The program lasted two months — an ice camp on the Soviet model, complete with a 1.6-kilometre airstrip, was set up at 86 degrees N, 110 degrees W, and operated from April 3 to May 20. "Long enough," said Peter, "to do 36 piston cores on the Alpha Ridge. What we found were Campanian and Maastrichtian sediments, very late Cretaceous, which suggested that the ridge itself had been formed somewhat before that, say during the mid-Cretaceous." These tallied nicely with Art's 1988 findings. But how did the ridge get there?

"We think it might be a hot-spot spreading center," said Art. A hot spot is a thin layer of crust above an undersea volcano; as the volcano bubbles up through cracks in the crust, a pile of hardened lava forms. Eventually this pile rises to the surface of the ocean and is called an island. Iceland is a hot spot. Hawaii is a hot spot. "If the Alpha Ridge is a hot spot, then it could be the same one that went on later to form Iceland and part of Greenland. We need a seismic line and more piston cores to see if the Alpha is continuous with those or a separate phenomenon. It could be very exciting."

As the *Nathaniel B. Palmer* made its way toward Antarctica, I recalled reading, Barry Lopez had stood on the main deck and looked out over the vast ice field, then put his ear to the ship's rail and listened to the distant throb of the *Palmer*'s powerful engines, pulsating up from several decks below. On the *Polar Sea*, he could have heard the same thing by going to bed and putting his head on a pillow. I tried sleeping with my ear over the gap between the pillow and the mattress, on the theory that the dead air would muffle the sound, but instead the cavity seemed to act as an amplifier. I felt as though I was lying with my head on a railway track, and a train was coming. When I eventually fell asleep, wearing a pair of ear protectors that Malcolm had borrowed from the *Louis*'s engine room, I dreamed I was being carried on a powerful river current, surrounded by large and incongruous items, all of us floating out together into a wide, turbulent, radioactive sea.

I woke up at 2 A.M. with a piercing headache, which I attributed to severe caffeine withdrawal. The coffee on the *Polar Sea* was so weak it barely deserved to be called coffee. Either it was being rationed severely, or else it was intended for 21-year-olds who hadn't yet become addicted to it. I had a theory that the coffee plant is actually a sentient parasite that has chosen human beings as its host: our addiction compels us to cultivate the plant carefully, pruning and watering and fertilizing and grafting and pampering, thus ensuring the comfortable proliferation of the species. If it were up to the *Polar Sea*, however, the coffee plant would become extinct.

I decided to get up, drink a pot or two of coffee, and see what was going on above decks. Most of the scientists had gone to bed, but there was usually someone in the crew's mess, drinking grape juice or playing chess. Tonight the mess and the coffeepot were empty. The galley staff had left a stack of pre-filled filters beside the coffeemaker. I dumped the contents of three filters into one and made half a pot of tolerably strong coffee. If the ship ran out of coffee three pots before Point Barrow, I figured no one but me would notice. I took two paper cups of coffee into the xo's office and read for a while, then, with two more cups, I went up to the officers' lounge. One of the engine room officers was watching *Blazing Saddles* on television and eating microwaved popcorn. I sat in the chair beside him and rubbed my eyes.

"Headache?" he asked.

"Second night in a row," I told him. "I think it's caffeine withdrawal."

"Naw, it's the ventilation system," he said. "It's a real dinosaur. They don't even make parts for it any more. Dries out the air. After a day, your sinuses dry up; after two days, your nose starts to bleed; after three days, your brain starts to shrink."

I didn't believe him, but I inhaled the steam rising from my coffee, just in case, then asked him about the *Polar Sea*'s fuel capacity. There had been rumors that Captain Brigham was so worried about running out of fuel that he was talking about turning back. This struck me as somewhat irrational, since we were nearing the spot in the ice cap known as the Pole of Relative Inaccessibility — the point equidistant from the ice edge. Maybe he thought that if he was going to run out of fuel, it would be better to do so as close to Alaska as possible. Eddy had told me that Captain Grandy had volunteered to transfer a week's supply of fuel to the *Polar Sea* if she ran low, and David Johns had even offered to send a resupply ship to Point Barrow to meet us.

"Well," said the officer, "you have to remember that this is a new situation for us up here. I mean, we're used to the Antarctic, where you go through no ice for a long time, and then a lot of ice for a few days. Up here, we're going through so-so ice but for a protracted period, so we don't really know how much fuel we'll need. We set out with 1.36 million gallons [6.18 million litres], about a quarter of what the *Louis* holds, and with all six engines going we'd run out of that in 38 days. We've been in the ice, what, 18 days now, but we haven't been at full power all that time, so depending on who's doing the math you can be worried or you can say we've got lots of fuel left."

"That's without using the gas turbines," I said. "What happens if you have to turn those on?"

"Well, they really suck up the juice, no question. If we have to turn those babies on, we're either outta here in ten days or we're here for the winter."

Over the next three days we did four science stations down the eastern slope of the Mendeleyev Ridge, moving out over the Makarov Basin, the 3,000-metre-deep valley between the Mendeleyev and

Lomonosov Ridges. Knut had decided, in consultation with Art, Eddy, and the two captains, to run for 18 hours a day and then stop for a six-hour science station every night. This would save fuel by allowing the two ships to sail in tandem, one breaking a trail for the other. It would also make for stations that were long enough to allow ice work, engine overhauls, and ice reconnaissance flights.

"It means that where we do the stations will be determined by where we are at 10 P.M.," Art told us at our regular science meeting on the evening of the 9th, "rather than by specific latitudes and longitudes. So we can't predict where we'll be for any given station. But we're short two things — time and fuel. This schedule will save fuel, and I think it'll save time. Right now the ice is still melting, but that will change in a few weeks: we don't want to be here when it starts to freeze up again."

The next morning, the temperature was 0.6 degrees C, and the air was full of fog and drizzle. I was on the bridge when the *Polar Sea* took over the lead from the *Louis*, and I watched the Canadian ship disappear into the dense fog in our wake. Within minutes we were alone in a huge, empty cloister of wet cloud. Rain oozed against the bridge's windows and ran down the panes like quicksilver. On the *Louis*, the captain always steered the ship from the bridge, 17 metres above the ice at eye level. When the *Polar Sea* was in ice, the ice pilot worked from the "aloft con," or crow's nest, a small, cramped, instrument-filled compartment at the top of the mainmast, 32.3 metres above the waterline. I never saw the sense in this, but it was a Coast Guard tradition, developed for Antarctic conditions. The xo thought the tradition went back to 1956, when the deck officer on the uss *Glacier* couldn't see far enough ahead from the bridge, so he went up in the ship's helicopter and radioed directions to his helmsman. But placing the pilot in the shrouds was an old whaling trick: George Washington De Long sent his ice pilot up to perch on the *Jeannette*'s topsail yardarm when he entered the ice off Herald Island in 1879. The *Polar Sea*'s aloft con contained its own set of control levers that communicated directly with the engine room; when Steve Wheeler, the ice pilot, was in the aloft con, no one needed to be on the bridge at all, although someone always was.

At the moment there were five of us: three Coast Guard officers, Art, and myself. The officers were looking aft, trying to find the *Louis* in the haze. Every now and then the deck would shudder loudly as a particularly large chunk of ice milled through the propellers. Art and I were watching the ice loom up at us as the prow rose, then slip away beneath our feet. We were inching through a loud and grizzled world, a sea of gray fog, white ice, and black water; it was like disappearing into the screen of an old black-and-white television documentary with the volume on full.

"We're into yesterday now," said Art. "We crossed the international date line last night. This is now yesterday. Weather seems about the same as it did tomorrow."

"Another day lost," I ventured.

"It's just time," he said, smiling like a Buddha. "A friend of mine was up here last year in a navy submarine, doing gravimetric measurements in the Canada Basin as part of a heat flow experiment. Now there's someone who knows what it's like to have a program canceled. In 1969, he'd had an experiment accepted by NASA for the first moon landing. Can you imagine his excitement? He spent years setting it up. Then those clowns in space suits started fooling around with golf clubs for the camera, and one of them kicked a cable and ruined his entire experiment. He doesn't like to talk about it yet."

The *Louis* beat her own record — 82 degrees 47 minutes N — on August 11, at 12:15 P.M., while Stefan and I were still on the *Polar Sea*. Captain Grandy had taken a ship farther north than any other Canadian captain in history. I wrote a congratulatory message and faxed it to him, and on my way back from the radio room I stopped in to see if Bob Writner was going down for lunch. I found him in his cubbyhole, staring at his computer screen; from the expression on his face I could tell that something unfathomable had gone wrong somewhere out in cyberspace.

"We've lost TeraScan," he said when I sat down. "We're not getting any more SSM/I images."

"What does that mean?"

"It means no more ice maps," he said. "And I can't figure out what's wrong. It could be software — it's a brand-new system, after all — or it could be hardware. I dunno."

I went down to the mess, where I found Jim St. John squeezed into a chair eating something square buried under a pile of creamed corn. I told him about the TeraScan problem.

"Yeah, I know," he said. "I heard from Caren. That's why I came over. I think it might be the antenna. It's programmed to follow the satellite's orbit as it passes over our heads every 110 minutes, but since nothing was ever intended to be used this far north, it was probably programmed to use magnetic compass readings. And of course, up here, magnetic readings are useless. Magnetic north is due south from here."

"If that's the case, how come it worked until today?"

"Dunno," he said. "Good question."

"What else do we have on board that wasn't intended to be used this far north?"

Jim laughed. "Everything," he said. "Inmarsat, *Marisat*, ERS-1, NOAA, *Healthsat*, these ships, my video camera, you, me. This is a whole new ball game up here, my friend. This is real seat-of-the-pants stuff."

"What about the GPS?" I asked. There are twenty global positioning satellites fixed in stationary orbit above the earth, and each ship had four GPS antennae that picked up signals from at least three satellites, making it possible for the ship's computer to triangulate its position to within a 100th of a second, or approximately 22 centimetres. Theoretically.

"It's working so far," said Jim. "But we're going to lose the others. I'll bet dollars to donuts that when we get to the North Pole, the only way we're going to know we're there is by the GPS."

"And the only way we're going to be able to tell anyone about it," I said, remembering Gord Stoodley, "is by Morse code."

That the *Polar Sea* was still in touch with the world-as-we-knew-it at all was thanks to the Lincoln Experimental Satellite (LES-9), a U.S.

Air Force device that linked the *Sea*'s shipboard computer to the University of Miami, in Malabar, Florida, through e-mail. It was still working, and would work right through the voyage. On August 12 we learned via Florida that Tony Gow had been awarded the Seligman Crystal, the Oscar of the International Glaciological Society, for his work on Greenland ice core dating, and that afternoon Captain Brigham held a small reception for him. Knut, Eddy, Caren, and Mary Williams flew over from the *Louis*, and all the scientists on the *Polar Sea* were there. The captain ordered cold cuts and cheese trays from the galley, and there was coffee (real coffee, brewed in the captain's galley) and Coca-Cola on a side table under Captain Brigham's pennant from the Explorers Society, which he had promised to hoist at the North Pole "should we be fortunate enough to make it."

It was like a reception in the staff room of a small but prestigious New England college. Tony was wearing a jacket and tie, and Captain Grandy was in his dress uniform, but the rest of us were more casually turned out. Knut made a brief congratulatory speech, then Terry gave another, more detailed speech, then Captain Brigham presented Tony with an official Coast Guard Certificate of Congratulations. On the coffee table was a large layer cake with sea-blue icing and a figure of the *Polar Sea* in red and white, and the words "Congratulations Tony" arched wavily above it, like northern lights. The captain cut the cake with his ceremonial Coast Guard dress sword, and then we all stood around holding small plates and glasses of Coca-Cola or, as in my case, cups of coffee.

I found myself next to John Grovenstein, an atmospheric physicist from the University of North Carolina, who was working on the very complicated subject of cloud formation. If global warming modelers had trouble including oceanic variants in their models, they had even more difficulty accounting for the influence of clouds. Do clouds block thermal radiation from reaching the earth, thereby inhibiting global warming, or do they keep it in once it gets here? Would surface evaporation create more clouds, or would desertification mean less? And if modelers were vague about how the Arctic Ocean affected global climate change, they threw up their collective hands over the question of clouds above the Arctic Ocean. John tended a large,

elaborate electronic gizmo on the bridge that counted cloud condensation nuclei (CCNs) — aerosols of various sizes and composition —
and assessed their relative effectiveness as cloud-forming particles by
injecting them into a computerized cloud formation program. The
idea was to see what kind of clouds the Arctic would form if it were to
form clouds at all.

"The problem," said John, who was a well-groomed, clean-cut,
entirely serious young man and the only scientist in the room except
for Tony wearing a tie, "is that Arctic air is so clean, I was having trouble getting results."

"That's too bad," I said.

"Yes, it is. I'm measuring anthropogenic pollutants up here in the
10-parts-per-litre range that back home in Raleigh I'd be getting in the
thousands."

"Good old North Carolina."

"Well, I know it sounds odd to be complaining about low pollution levels, but it really is a lot more complicated than you'd think.
Let me give you an example. You know how acid rain forms: sodium
dioxide from industrial smokestacks gets into the lower atmosphere,
mixes with water vapor and forms sulfuric acid, and then is precipi-
tated out onto lakes and forests. But before it rains down, it actually
blocks ultraviolet radiation from the sun, causing global cooling.
According to the models, Raleigh should be showing a 0.2-Celsius-
degree increase in average temperature by now; instead, it's showing
none at all. The hypothesis is that sulfuric acid in the clouds is creating a negative feedback."

"Do you mean to say that acid rain may save us from global
warming?"

"That's correct. How's that for irony?"

"I hope Inco doesn't hear about this."

Tony Gow also threw an interesting sidelight on the global warming issue. His work in Greenland, conducted in 1990, had been to extract a 3,200-metre core of ice from the ice cap, a 15-centimetre time
capsule with layers so sharply defined that he was able to date each
one precisely, year by year, down to 250,000 years before the present
at the bottom of the core. ("Actually," he said, "we brought up a bit of

bedrock from the bottom, so it goes back considerably farther than that, but it's the ice we were interested in.") Glacial ice is really compressed snow and air bubbles; in Greenland, it builds up in the manner of tree rings, trapping dust particles, acid molecules, and gases that allowed Tony to assess what was in the air when the ice formed.

"It's actually very interesting," he said, "to hold a piece of ice in your hand that was formed the year Shakespeare wrote *Hamlet*, or the year of the Magna Carta, or the year the Vikings settled Greenland."

He was also able to identify the decade, if not the year, that the last ice age began to end. "At 1,678 metres down," he said, "we came to a layer of ice that was formed exactly 11,680 years ago, and we could clearly see the beginning of a warming trend — the layers became thicker, some were curved. We saw the flickering switch, the end of the last ice age and the beginning of the Holocene. We saw global warming. The average temperature in Greenland suddenly rose 8 Celsius degrees, and that was enough to put an end to 1,300 years of continuous freezing."

"Suddenly?" I knew that geologists used words like "suddenly" to describe time frames like the end of the dinosaur era, which may have been spread out over 2 million years. "How long did it take for the average to go up 8 degrees?"

"Ten years," he said.

Our worst-case global warming scenarists predict a 5-degree rise in average global temperature by the end of the next century; here was Tony saying that the last ice age ended with an 8-degree rise in Greenland that was accomplished in a single decade. And that was without the aid of anthropogenic carbon dioxide, fluorocarbons, radionuclides, and all the other unnatural compounds that are arguably hastening our demise.

"Which means," he said, "that global warming is not the end of the world."

"Oh, I quite agree," said Evelyn Sherr, a marine biologist from Oregon State University, who had joined our conversation. "I don't think global warming is our most pressing problem. Even if all the predictions bear up, it'll take a hundred years to do us in. I think our immediate problems are overpopulation, deforestation, soil erosion, dwindling fuel reserves, and the end of our fresh-water supply. Global warming won't be the end of the world. The earth will survive as it has always survived. But these other problems could spell the end of our civilization. I think they are far more serious than a change of climate."

After the reception, Knut and Art held a brief meeting to discuss the science plan for the coming weeks. There were two ways to head north; one was to continue straight on to the Alpha Ridge and then make a left turn north to the Pole; this was termed the counterclockwise route. The other was to work our way northeast in a clockwise direction, then ride up the Lomonosov Ridge toward the Pole along 155 degrees E. We would then circle back south along 145 degrees W to the tip of the Alpha Ridge, when Art would get a chance to run his seismic lines on the return from the Pole. Knut favored the clockwise plan. "At this time of year, the ice gets better every day because it's still melting, so the later we approach the Alpha Ridge the better conditions we should have for seismic work."

Art, however, argued for the counterclockwise route because he wanted to get his seismic crew working sooner rather than later, which would be closer to freeze-up. But he agreed that ice conditions would determine the route. They decided to send a helicopter ahead the next day, to reconnoiter the clockwise route.

I had fallen into the habit of going down to the geology lab first thing every morning to check our position. Located on the same deck as the other labs, the XO's office, and the crew's mess, lounge, and exercise room, the lab was an island of calm in a maelstrom of activity. The geology computer had a neat little program that combined GPS data with meteorology and geology: it gave out a constant read of such useful information as our latitude and longitude, ambient air temperature, wind velocity and direction, and the depth of the ocean beneath us. In a box in the corner of the screen, we could enter the coordinates of our last science station and those of our next science station, if we knew them (or our sailing time to it if we didn't), and the visual display would show a cartoon of the two stations and a little Popeye ship puffing along a direct line between them, as well as the time of our arrival if we kept steaming at our present speed.

But the morning after Tony's reception, I found the lab in a state of mild panic. The computer had crashed. Fred Payne, Steve May, and Pat Hart were hovering around one terminal; Bill Robinson and a Coast Guard technician were staring at another; and a third monitor was blinking "System Error" at us in three attractive colors. No one could figure out what had happened. The program printed GPS data on the seismic reflection charts, so that the geologists could tell the exact coordinates of each science station and every sea-floor feature we passed over. They needed this information to determine where they would do their seismic tows and piston cores. Without the program, said Fred, "it'll be like coring in the dark. We won't know where the hell we are."

"It crashed at 84 degrees exactly," said Pat. "Just went blip."

"We were told something like this might happen," said Steve. "But we thought we'd probably only lose the graphics. We thought the numbers would still be okay."

"Maybe all we have to do is go in and reprogram it for above 84," Bill suggested.

"I don't understand why nobody tested it," said Steve. "All we had to do was punch in 85 or 90 degrees and see what would happen."

"Maybe somebody did but they didn't recompile."

"There's a thought. It'd be easy to find out."

Reg clicked away at the keyboard. The screen flashed: "File Error."

"I'm going to bed."

"Yeah, this thing can wait."

When everyone but Steve and Fred had left, Fred started morosely taking apart the mouse with a screwdriver. "Might as well clean the tracking ball," he said.

"You just don't know what's going to happen up here," said Steve. "No one knew how to prepare for this. When you boldly go where no one has gone before, you're gonna run into things no one ever dreamed possible. Steve Wheeler says he punched 90 degrees in the GPS the other day to see how far we were from the Pole, and the whole thing blew up on him. Our program here runs off GPS. If GPS goes down, whether we get our system up and running again is irrelevant."

"If GPS goes down," said Fred, pulling a nest of lint out of his mouse, "we might as well pack up and go home."

"Yeah," Steve said. "But which way will home be?"

14

Hydrometer Rising

The experienced oceanographer takes precautions to avoid any inconceivable equipment misbehavior and expects that highly improbable events will surely occur.

— H. W. MENARD, *Anatomy of an Expedition*

W E PASSED 85 degrees on August 14, enshrouded in fog. TeraScan was still down, the geology computer was still going haywire, the GPS was winking on and off, but we knew where we were. There was some flurry on the *Polar Sea*'s bridge, because 85 degrees was a new farthest north for an American surface ship. Steve Wheeler relayed Captain Brigham's congratulatory message to the crew over the ship's intercom.

The sun's pale disk, hovering just above the southern horizon, had emerged for an hour that morning, but the fog had long since defeated it. Standing on deck, I couldn't tell whether we were moving, let alone how fast. The temperature was −2.5 degrees C, which meant winter was closing in. Staring aft along our wake I fancied I could see the dark outline of the *Louis*, but I couldn't be sure. It might have been a trick of the fog, or a mountain of ice, or a sea monster plunging silently through its own steam. When we entered a fogless patch, it was like breaking into a clearing in a dense forest. The ice was tightly packed, with no leads or polynyas through which we could thread our way north. As we passed through one of these fogless spaces, the *Louis* sent up a helicopter for an ice recon, like Noah sending out a dove, but the helicopter didn't get more than 20 kilometres from the ship before turning back: too much fog. Without imagery from the satellites or helicopter recons, we were in a formless world, nosing our way through ice and into leads like blind mice in a maze. The GPS was still operating, so we knew where we were. But we had no idea how to get where we wanted to go.

We were still wavering between the clockwise and counterclockwise routes to the Pole. The helicopter pilots bravely made two more ice recon attempts, but both were turned back by fog. I was supposed to go back to the *Louis* that night, but I didn't mind much when my flight was canceled because of the fog. I had grown to like the *Polar Sea*. Its blatant militarism grated on me, perhaps because my father had been in the air force and I had grown up on military bases; but beneath the swagger was a laxness that, if you knew how to take advantage of it, allowed everyone who kept his or her head down to just get on with things. It was a Sergeant Bilko kind of militarism. I had been aboard the *Polar Sea* long enough to see through the posturing and the machismo to the core of civility. Perhaps it was the fog pressing against the bulkheads that made the softly lit interior seem cozier, like a log cabin in a snowstorm. Or maybe I had survived caffeine withdrawal and was now less surly. For whatever reason, I was more attuned to the nuances of life aboard the *Polar Sea*. I had the run of the ship, the scientists had lots of work and seemed content despite our lack of forward progress, and I enjoyed the loud rhythms and camp atmosphere that prevailed.

That night, we stopped for station 28, well out over the Siberian Abyssal Plain — at 3,900 metres, the deepest part of the Makarov Basin. It would be a long station. We were in a small polynya surrounded by expanses of soft, undulating white, like a farmer's field in winter. It was a pleasant evening for a stroll, and I walked out onto the ice with Captain Brigham. We conducted a sort of royal tour of the various areas of scientific activity, nodding congenially and asking pertinent questions. Terry and Erk had found a small field of dirty ice. Michel Gosselin was collecting copepods. And the *Polar Sea*'s scuba divers were helping out the biology team by scooping up great bucketfuls of ice algae (*Melosira arctica*) and plucking amphipods from the underside of the ice. Pat Wheeler was working on a theory that ice algae were performing many of the duties normally assigned to large-celled phytoplankton, and wanted to ask Chris what the algae's iron requirements might be. Everyone seemed to be doing basic biology. We walked from one knot of activity to another, the huge red-and-white ship forming a backdrop to the illustrative scientific panorama.

Captain Brigham, in a philosophical mood, returned to a favorite theme — the degree to which the U.S. Coast Guard was becoming a civilian organization. He had clearly enjoyed his four years at Coast Guard headquarters in Washington, whereas Captain Grandy couldn't wait for his 12-month stint at a desk in Ottawa to end.

"Our role has been as much civilian as military right from the start," said Captain Brigham, "at least up here in the North. We've been as much a supply service to northern outposts as a defense against invasion. It was a Coast Guard ship, the *Corwin*, that brought John Muir to Alaska in 1881, remember. There was a science component on the search for the *Jeannette*. There were scientists, filmmakers, and even a writer on the *Northwind*. And as I've said, I see my own role on the *Polar Sea*, at least on this operation, more as a coordinator of the different communities on board, a kind of chairman of the board, than as a captain of a ship that might be called into battle."

I asked if that was the way he saw the Coast Guard evolving.

"Well, it's hard to say," he admitted. "Most of the Coast Guard fleet is in the Gulf of Mexico, keeping Haitians out of Florida, so it's difficult to maintain that we're turning more and more of our ships into science platforms." During his Washington years he had written an article for the Coast Guard publication *Proceedings* in which he detailed the complexity and diversity of the Coast Guard's role. The *Rush* had raced to the Marshall Islands in February 1993 to quell "a major smuggling operation involving 500 Chinese migrants on board the MV *East Wood*." Another Coast Guard cutter had worked on an oil pipeline rupture in the Potomac River. Another had helped clean up a 3-million-litre oil spill off the coast of Puerto Rico. Science support was a small part of the Coast Guard's overall picture. "The icebreakers, certainly," he said, "but icebreakers aren't a big concern except on the Great Lakes. Within the Coast Guard's entire budget of $3.67 billion, we're peanuts. Our polar operations are about 3.6 percent of that. Compared with law enforcement, we could disappear tomorrow and no one would even notice."

I had heard Captain Brigham described as an ambitious man. "He wants to be the next commandant of the Coast Guard," Grandy had said. But what I had just heard was not the calculated statement of an ambitious man, nor even the satisfaction of a sea captain who had just set a new record for his country's farthest North. He seemed tired, even discouraged. I asked him about his plans when he returned from this expedition.

"I'm going to take early retirement in the spring," he said. "I've applied to the Scott Polar Research Institute, in Cambridge, England. If I'm accepted, I'm going to go there to get my PhD in remote sensing, and then I'm going to go somewhere and teach."

We were standing near the *Polar Sea*, admiring the line of her bow as it curved down into the dark blue Arctic water. From the bridge's side window, one of his officers leaned out and waved; Captain Brigham waved back, then made a slushy snowball and threw it as hard as he could at the window.

Back on the *Louis*, in the officers' lounge, I was drinking Scotch and playing chess with Doug Sieberg. Despite my new fondness for the *Polar Sea*, the *Louis* still felt like home. Mary Williams was watching the game and talking about the Confederation Bridge, Prince Edward Island's projected "fixed link" with New Brunswick, scheduled for completion in June 1997. "There have been dozens of studies into what kind of ice the bridge is likely to see," she said. "I've read them; all they talk about is how much and how thick. They're all based on an extraordinarily static idea of ice. They think ice just sits there and grows. Nobody seems to appreciate that ice is moving all the time, rafting, shifting, drifting. The total volume of ice flowing through Northumberland Strait and likely to hit the pylons of a bridge is way out of whack with their models."

"Check," said Doug. I intruded my bishop between my king and his queen.

"It's the same in the Arctic," Mary went on. "All these floes are composites, each one an amalgamation of ice of different ages, different origins, different sediment content, different temperatures, all stuck together in one so-called floe. No single concept of ice can account for all that. No single ice map could prepare you for it. I don't know, it's almost as if you have to get out and walk ahead of the ship, like the Inuit do with their dogsleds."

"Check," said Doug again. By moving my bishop, I had opened a lead between his rook and my king. Thinking more about Mary's view of ice than about chess, I moved the king into what looked like a safe harbor behind a staggered phalanx of pawns. Doug frowned.

"We have to start thinking of ice as a living organism," Mary said. "It is not static; it is not rock. It's more like soil. We have to approach ice as biologists, not as geologists. There is an amazing amount of biological activity going on in and around the ice. You can't think of ice as inert skin growing on top of the water. It's actually a total environment, a biological system. It has plant life — algae, phytoplankton, diatoms, microbes — and it has animal life, everything from zooplankton and copepods to seals and polar bears."

Doug was too polite to say anything as pointed as "Checkmate." What he said was, "Would you like a cup of real coffee? We have some fresh beans and a Melita coffeemaker in our room."

"*Louis St. Laurent, Louis St. Laurent,* Delta Uniform Four-Three-Niner. Come in, please. Over."

Captain Grandy picked up the radio handset that hung from a hook on the bridge's starboard bulkhead and spoke into it. "Yes, Delta Uniform Four-Three-Niner, this is *Louis St. Laurent.* Go ahead. Over."

"What can we do for you today, Captain? Over."

"Thank you, Four-Three-Niner. I'd like to know the ice conditions between 89 degrees north, 149 degrees east, and the Pole," said Captain Grandy. "And then on your southern leg I'd like the same information from the Pole to 89 degrees north, 150 degrees west. Over."

I was amazed and a little disappointed that it could be this simple. Delta Uniform Four-Three-Nine (I couldn't bring myself to say "niner") was an Aurora aircraft belonging to the Canadian Maritime Air Command. To make a more informed decision about our ultimate route to the Pole, clockwise or counterclockwise, Captain Grandy had requested an overflight along both possible alternatives. The airplane carried an Ice Patrol observer and a Coast Guard captain and was making a visual assessment of both routes. We were still south of 88 degrees, so the distance to the Pole along either route was well out of our helicopter range, and with TeraScan down, there was no way of getting the information other than sending Ikaksak ahead with a snowmobile. Still, picking up the phone and calling in the armed forces seemed a trifle anticlimactic.

"Uh, that's a roger, Captain. No problem. We are now approximately one hour from your position. And we have someone on board here, Dr. René Ramseier, who requests that Caren Garrity be on the bridge during our fly-by, if that's possible. Over."

"Roger, Delta Uniform Four-Three-Niner. That's a roger. I'll pipe Dr. Garrity to the bridge. Over and out."

Forty-five minutes later, everyone, scientists and off-duty crew members, was either on the bridge or outside on the decks, plane spotting. It was another gray day streaked with dark, windswept snow. The ship's engines and the sound of wind in the cables at first obscured the distant thrum of the plane. The Aurora was a camouflage-green, four-propped aircraft, strangely old-fashioned; its approach through the fog was like a scene out of *Casablanca*. I was on the roof of the helicopter hangar with Alethia Lee, the ship's nurse, and Mark Cusack, the chief engineer, trying to keep the snow off my binoculars. The plane made a pass over the ship about 60 metres above our heads. It then carved a wide turn astern of us, swinging back around to our starboard; another wide turn, another circle, and this time a small canister dropped from its belly onto the ice about a kilometre to starboard. This was a dummy canister, dropped to give us practice in retrieving the real thing when the plane returned from its reconnaissance flight, when the canister would contain a hand-drawn ice map that would tell us where, if anywhere, the ice was easiest. Steve and Mike took off from the flight deck in the *Louis's* helicopter and retrieved the dummy canister. Steve brought the canister up to the bridge.

I went in with him. It was very crowded, and everyone was listening to Caren talking to René on the ship's radio. "I'm glad to see you're keeping busy during my absence," she said. "It makes me feel better about being here on the ship. Over."

"Roger," said René, obviously aware that he was being heard by 35 scientists and several dozen crew members, as well as by his wife. "It's good to hear your voice, Caren. When we drop the real canister, there will be a letter in it for you giving more details." Caren and several dozen of us on the bridge smiled.

"My God," Brenda said in a loud whisper. "Are we all so starved for contact with the rest of the world that we have to eavesdrop on Caren's conversation with her husband?"

"You bet we are," said Steve. "And if I get to that canister first, I'll probably read her letter, too."

When the Aurora returned two hours later, Fred brought the second canister up to the bridge. It was a long brown tube with "Courtesy 407 SQN, Crew 4, Icemen" written on the side in black marker. Fred cut off the top with his pocketknife and pulled out the contents, which consisted of two rolls of toilet paper (presumably as packing), Caren's letter (unopened), a stuffed gerbil, a copy of the previous day's *Victoria Times*, and two hand-drawn ice maps of the alternate routes to the North Pole. Fred took the maps and gave the newspaper to Captain Grandy, who took it apart section by section and began auctioning it off.

"Who'll give me a dollar for the comics?" he said.

Fred and Caren's reading of the ice maps told them that the ice was pretty much constant between our position and the North Pole. If you looked at the two longitudinal routes, 149 degrees E and 150 degrees W, as two sides of the letter A, with the Pole at the top, it didn't much matter which side we took. The problem was that we were currently at the base of the left-hand side; to get to Art's route, we would have to cross over to the right-hand side, and the maps did not give us any information about what the ice was like in the crossover. Captain Grandy shook his head. The final decision would be made in consultation with the three chief scientists and Captain Brigham, but it was clear which way he was leaning. "We have to take the route with the least ice," he said, "scientists or no scientists."

Eddy had been hoping the overflight would show that one route was clearly easier than the other. It was not information but lack of information that favored the clockwise route, and if there's anything a scientist hates, it's lack of information. Eddy agreed with Captain Grandy's assessment. It meant yet another postponement of the seismic program, but we couldn't afford to get stuck in unknown ice. We couldn't put all the science programs at risk in a possibly futile attempt to implement one of them.

"Besides," Eddy said, "I think it's very important that we get to the Pole. We have to show the world that we can go anywhere we want up here, not just to within 100 kilometres of where we want." We looked down at the chart, where the *Louis*'s track line was penciled hourly on the plain white surface; the X that marked 90 degrees N was only a few centimetres from the last entry. "We're 120 nautical miles from the Pole. If we were in a car, we could be there in two hours. We can do the science on the way back."

That afternoon, Knut called a meeting on the *Polar Sea* to discuss the alternatives again. It was a lively meeting. Art argued for the counterclockwise route, and Captain Grandy favored the clockwise. Knut, sympathetic to both sides, found himself caught in a crossfire. "After two hours," he told me afterward, "we achieved unanimity. We would go up the Lomonosov Ridge on the Canadian side, proceed to the Pole, and then do the seismic tow over the Alpha Ridge on the southbound run with more efficiency." This was the clockwise route. The seismic program was again postponed.

The *Polar Sea*, in the lead, spent the rest of the day bashing against a huge floe of rafted multi-year ice that was 5 metres thick; in nine hours, she made barely half a ship-length of headway. Steve Wheeler, at the helm, made run after run at the ice face, running the ship's bow up onto the floe, then backing into his own wake to gather momentum for another charge. We watched from the *Louis*'s bridge, engines idling. "When the ice says no," Knut said, "it means no. I have no doubt now that our choice of route was the right one. We can't go east from here."

Meanwhile, Ron Perkin came back from one of his satellite stations with exciting news. His series of mini science stations, peppered across the ship's track, fanned out from the ship like a harrow behind a tractor. While the CTD crew on the *Louis* had conducted 30 casts so far, Ron had just come back from his 101st. All he had was a small, cylindrical CTD, not an entire rosette, and he didn't collect any water samples. But even a CTD takes time. He had to drill a hole in the ice, attach the CTD to a steel cable that ran off a small homemade winch run by a modified chain-saw motor, and lower the CTD to a depth of 2,000 metres, at the rate of 1 metre per second, turn it on, and then raise it at the same rate, collecting temperature and salinity data on the upcast.

"Today," he announced quietly, "I think I've discovered a new sea mount."

The seabed of the Canadian Basin is imperfectly known, but its main features are agreed upon. There are the three main mountain ranges — the Lomonosov Ridge, the Alpha Ridge, and the Mendeleyev Ridge. Between them are the basins — the Makarov and the Canada Basin. Sprinkled among these massive features are countless smaller geological entities, sudden abysses, a ripple of foothills, unexplained mesas, and tall, narrow, current-scoured pinnacles of rock called sea mounts. Ron had just found a new one.

"It was very foggy," he said. "We flew out to the station and set up the winch. I decided to let the CTD go down at 2 metres per second instead of 1, because the fog was closing in and I thought we should hurry and I also thought we were over deep water." According to the bathymetric chart, he should have been over water 3,700 metres deep. "All of a sudden, as the line was going down, it just sort of went slack at 850 metres. I thought it had broken — Damn, I thought, the CTD is lost. But then Oolat picked up the slack line in his hands and pulled on it, and said, 'No, the instrument is still there. You've hit bottom.'"

Ron realized at once that he had hit an undiscovered feature. Scientist that he is, however, he experimented around before breaking open the champagne. He and Steve and Oolatita flew to three other locations, all within a nautical mile of the first, drilled new holes in the ice, and lowered the CTD. Each time it went all the way down to the end of the cable. There was no doubt about it: Ron had discovered a sea mount — and an incredibly large one, one that rose nearly 3,000 metres from the ocean floor. He had just discovered an underwater mountain the size of Mount Baker.

For the rest of that day and the next, the ice remained forbidding, a uniform field of broken and refrozen ice stretching farther than the eye could see. That night marked the ice's birthday, the day first-year ice becomes second-year, and second-year ice becomes multi-year. It's the day the ice begins to refreeze, to grow thicker, its melt ponds coated with a thin layer of ice and camouflaged under drifting snow. Normally, ice in the High Arctic isn't supposed to refreeze until the middle of September; here it was the third week in August and already winter was upon us. Everything seemed to confirm the wisdom of Knut's decision to take the nearest route north.

Visually, we had progressed from spring to winter without an intervening summer or fall. The ice reminded me of René's discovery that refrozen ice is harder than ice in natural solid sheets; this jumble of refrozen monoliths looked harder to break through than the smooth floes we had navigated earlier. We were barely making headway; the bridge's monitor showed us making half a knot, and even that seemed optimistic. It was one ship-length forward, half a ship-length back.

"Wind speed," said Caren, "44 knots." We were at station 31. Caren, on her knees in the snow, wearing a bright yellow Mustang suit, had carved a tiny cliff into the edge of a drift and was measuring the temperature gradient from the hard, wind-stiff layer on top down through to the crushed glass at the ice surface.

"Ambient air," she continued, "0.7 degrees. Many things affect the satellite reading of snow," she added as she calibrated the resometer and jotted figures in a dog-eared and water-soaked notebook. "Heavy new snow obscures the old crystalline layer underneath, but if it's dry on top you'll still get a reading from the underlayer. Snow will have a higher albedo on a sunny day than on a foggy or cloudy day, because there will be more meltwater in it."

"How's René?" I asked as we packed up the resometer. She had told me that René had been under some pressure to take early retirement from the AES. A new director, different departmental goals, government cutbacks; the old story, or rather the new story for many scientists who have spent most of their adult lives working toward some kind of definitive scientific statement, only to have their support system yanked out from under them. In the U.S., the Reagan administration strangled global warming research in its infancy because its findings might have been antithetical to uninhibited industrial growth — which is why serious concern about global warming didn't start appearing until 1989, even though the early warnings had come in the 1970s. In Canada, the government seemed to be getting out of the raw research business, preferring to stick to proven technologies that delivered data it could sell; when *Radarsat* was launched, it wasn't going to have a passive microwave radiometer on it.

"He's decided to accept their offer," she said. "They came in last week and took away all our files and computers." We trudged against the wind to the edge of an ice-covered melt pond; I could feel icicles uniting my mustache and beard. Ahead of us, the ice field was spectacular, gentle undulations of snow-covered ice, long scarves of snow among the dunes, and above, in a sky studded with purple and rose-colored clouds, a thin sun, higher than usual, fighting to push its pale, warmthless light down to us. We were creatures without shadows.

"Kick a hole in that melt pond, would you?" she said.

"Gladly." I brought my boot down heavily on the pond's thin crust, and had a sudden memory of testing the ice on a winter pond near my boyhood home in Barrie, Ontario, to see if it was hard enough for hockey. Even my leg muscles remembered, for I brought my foot back up with a clean jerk before the water below could flood into my boot. "He's had a few offers from private companies already," Caren added as she measured the thickness of the crust and the depth of the pond. "Still, it makes me wonder why I bother doing this."

Monday, August 19: the exact midpoint of the expedition for the scientists, assuming that our original schedule — Alaska to the North Pole and back to Alaska — was still intact. It was known as hump day; all downhill from here. No one quite knew if it was the midpoint, of course, but we congratulated one another anyway. Both ships had been stopped since noon. I had been out on the ice again, this time with Chris Measures, collecting snow samples to test the dust in them for iron and aluminum, and with Mary Williams and Ikaksak, cutting ice beams with Mary's Husqvarna chain saw. The day before, Mary had been unable to start the chain saw. Oolatita, who had never seen a chain saw in his life before, took it apart, found the problem, repaired it, and put it back together again in an hour. "Where I live," he'd said, "if you can't take a two-stroke engine apart in the dark, you don't eat for six months a year."

In the lounge that night, with the ships still stopped, there was a feeling of elation. We were at 88 degrees 47 minutes N — 73 nautical miles from the Pole — and the sun was shining. Chris passed around a tin of macadamia nuts that his wife had hidden in his kit bag. That afternoon, Doug Sieberg and Louise Adamson had announced that they were going to be married by Captain Grandy at the North Pole, and now everyone was trying to buy them drinks.

They had told Captain Grandy about their plans before leaving Victoria, and although ships' captains are supposed to be able to marry people at sea (at least in international waters), Captain Grandy had taken the precaution of becoming a temporary justice of the peace for the Northwest Territories, just to make sure the wedding was legal. The official permission had arrived by e-mail only a few days ago, which is why Doug and Louise had waited to make their announcement. Canada was going to have its first wedding at the North Pole, conducted by a bona fide Canadian justice of the peace.

There was a ping from the ship's intercom. "Attention, please, this is the captain speaking." Captain Grandy had one of those voices that sounded better over the loudspeaker than they did in person. The conversation in the lounge subsided. "I have here our schedule for the next few days, which I would like to share with you. As we go down off the Lomonosov Ridge, we will be making station stops at 2,000 metres, 3,000 metres, and 4,000 metres. After that we will be making a 45-mile transit to the North Pole. Each station will be a two-and-a-half- to four-hour stop, and with a good transit we should reach the Pole sometime on Monday. Thank you."

"Looks like Monday's the big day, Doug," said Eddy, who was to be best man.

"Where are you planning to go for your honeymoon?" asked Robie.

"I'll build you an igloo," said Ikaksak.

A few minutes before midnight, we reached the 2,000-metre station, which marked the start of our passage over the Eurasian slope of the Lomonosov Ridge. I went out onto the ice with Malcolm and Robie. We lay down at the edge of a floe and hung out over the water to catch amphipods in a small dime-store fishnet that Malcolm had tied to the end of a 2.5-metre pole. Malcolm was becoming fascinated by these little creatures. One of the curious aspects of the large net hauls being done on both ships had been the low numbers of copepods in them. Copepods form nearly 70 percent of the zooplankton biomass in the rest of the world ocean: they are the principal food of the ocean carnivores, from the tiniest arctic cod to the bowhead whale. But there weren't a lot of copepods up here; there were fish and even whales, but they couldn't have been surviving on copepods. Their absence was tied in to the low-production conundrum that had plagued us throughout the trip: lots of nutrients, not much phytoplankton, few copepods, then lots of fish, seals, whales, and bears. There seemed to be a gap in the food chain between the bacteria and the fish. What was filling it?

Malcolm and Michel Gosselin, from the Université du Québec at Rimouski, thought it might be amphipods. Michel, working with the

Polar Sea's biology program, was interested in ice algae, the various species of *Melosira* that grow in the ice itself and in long, hairlike mats attached to the bottom of the floes. When ice algae die or are eaten by zooplankton, they give off a sulfur-containing gas called dimethylsulfoniopropionate (DMSP); in sea water, DMSP breaks down to produce dimethylsulfide (DMS), about 10 percent of which is released into the air. Once in the atmosphere, the sulfur in DMS influences the formation of clouds — hence the importance of the ice algae study on an expedition studying global warming. Michel wanted to know what kinds of ice algae there were and how much of them was being consumed, so he could calculate how much DMS they were contributing to the Arctic atmosphere.

In the process, he had begun to wonder whether the traditional phytoplankton-copepod links in the ocean food chain were replaced in the Arctic by a chain comprising ice algae and amphipods. Were these what filled the gap between nutrients and fish? Malcolm was interested in this possible alternative food web because he thought it might explain why the polar bears he was sampling contained lower levels of contaminants than polar bears in Svalbard and other nearshore environments. All the contaminants he was checking for — PCBs, DDT, the chlordanes, the whole alphabet soup — were turning up at about half the levels found in other polar bear populations. For some reason, high levels of toxic chemicals were not accumulating in the bears' adipose tissues: maybe a food chain beginning with ice algae and amphipods contained lower levels of contaminants than one that started with phytoplankton and copepods.

Even when lying on ice looking down at tiny organisms clinging to the sheer blue surface of the floe, Malcolm was thinking about polar bears.

When we returned to the ship I ran into Mary, who had just been talking to Eddy. The *Polar Sea*, she said, had developed "an alarming shudder." There are few things that can cause a 13,000-ton ship to shudder even mildly, let alone alarmingly. Icebreakers occasionally shudder when a chunk of ice, broken off by the bow, slides along the side of the ship and gets caught in the propeller and squeezed between the prop blades and the hull; the ship shudders for a few seconds as the ice is augered through the blades, a phenomenon known as milling. But that doesn't happen often, and it isn't alarming when it does. Unless the ice is so hard that it jams in there, in which case it might break off a propeller blade, or even bend the propeller shaft. Something like that might cause a ship to shudder continually, and that would be alarming.

Eddy came into the mess looking fatigued and confirmed that something was wrong with the *Polar Sea*'s starboard propeller. The *Sea*'s divers would go down to assess the damage.

"It isn't crippling, so far," Eddy said, "but it's worrisome. Anything that isn't right up here is worrisome. It'll be tough sledding these last 45 miles to the Pole, and Captain Brigham wants to make sure the *Sea* is in good shape."

"The *Polar Sea*'s propeller problems are not alarming," Knut began the next day's science meeting, which was rather alarming. "But they are something to be aware of. It seems her propeller took a real hard hit milling ice yesterday, and the retainer plate that holds her shaft bearings in place has apparently sheared off. The divers are taking another look at it now. But we've come 1,658 nautical miles through the toughest ice in the world, and we're not even halfway home yet. We have to take it easy."

I looked at Caren. She had been on the *Polar Star* in 1991, when a similar propeller problem had forced the ship to drop out of the *Oden-Polarstern* expedition and return home. With only two functioning propellers, the *Polar Sea*'s icebreaking ability was cut by a third. Underpowered to begin with using only her diesel engines, she might now have to resort to her gas turbines, and that would quickly get her into fuel difficulties. She might soon find herself without the horsepower to go forward, and not having the fuel to go back.

Eddy outlined the rest of our course trajectory. If you could drain the water out of the Arctic Ocean down to the level of the Lomonosov Ridge, you would see that the ridge takes a little jag at the North Pole, as if when the cartographer stuck the point of his compass there, the ridge had squirmed a little to get out of its way. We were over that jag now: beneath us was the deepest abyss in the entire Arctic Ocean, a straight drop of 4,308 metres.

The other thing you might see is a small notch in the ridge, just to the south of the Pole toward Canada. It's a hypothetical notch, deduced from slight temperature variations between Atlantic water on the Eurasian side and Atlantic water on the Canadian side, and a decided increase in the velocity of the Atlantic Layer. The theory is that Atlantic water spills through the notch the way wind blows through a mountain pass. Knut didn't think there was a gap there; the hypothesis was based on a few measurements made from the *Oden* in 1991, and no one had done any real coring or depth metering in the area. He and Jim Swift had checked the *Oden* figures again and again during this trip, and they couldn't conjure up a gap from them. But Knut wanted to collect some hard data of his own. "I've been so tied up with being the expedition leader that I haven't had time to look at my own particular interest, ocean circulation," he said. This was to be his chance.

That morning he and Art had flown a 215-nautical-mile helicopter recon of all possible routes to and from the Pole, and both agreed that coming down off the Lomonosov Ridge at 150 degrees W was impracticable. We would have to get to the Alpha Ridge another way: by doubling back on the Eurasian side and cutting across the hypothetical gap, dropping box and piston cores along the way, and thus approaching the tip of the Alpha Ridge, where Art could do his seismic tow. It was another alteration in plans: we were still taking the clockwise route to the Pole, but we would revert to a counterclockwise route after leaving it.

We finished the 4,000-metre station, at the bottom of the Lomonosov Ridge on the Eurasian side, at 11 P.M. on August 20, and stayed put until 5 o'clock the next morning while the *Polar Sea*'s engineers and divers tried to assess the damage to their starboard propeller.

Despite Eddy's and Knut's concern over the *Polar Sea*'s predicament, both were excited by the CTD readings. The warm-water layer was not a bolus or an eddy, it was a reality. The four stations we had just completed showed not only that the anomaly extended right across the ocean, but also that the water in it was becoming warmer — much warmer. Even Jim was convinced: he had been plotting the temperature curves after each station, and his excitement grew hourly as he compared this new data with Eddy's 1993 findings. The 50-metre layer was half a degree warmer than it had been in 1993, which made it a full degree warmer than it had been in 1991. And it was a lot more spread out than anyone had dared to imagine. Warm water was now in the core of the Atlantic Layer in three basins — the Nansen, the Amundsen, and the Makarov.

Eddy also now knew that the warm layer was rising: the rosette casts were clearly showing that its midpoint had risen from 400 metres to 200 metres in slightly more than a year. And it was displacing the cold, Pacific-origin waters of the halocline that usually occupied the 100- to 200-metre zone, which meant it was removing one of the main buffers that had been keeping it away from the surface. Something big was going on.

15

Ultima Thule

In place of the adventurous explorers of Frobisher's day, searching for fabled empires and golden cities, there appeared in the seas of the north the inquisitive man of science, eagerly examining the phenomenon of sea and sky, to add to the stock of human knowledge. Very naturally, there grew up under such conditions an increasing desire to reach the Pole itself. . . .

— STEPHEN LEACOCK, *Adventures of the Far North*

O N SUNDAY MORNING, Eddy wrote his own version of Heisenberg's uncertainty principle on the lab's bulletin board. Heisenberg had stated that it is impossible to determine simultaneously both the position of a body and its momentum, because the more you knew about where a body was, the less you knew about how fast it was moving. Eddy's version, which he called Carmack's Arctic uncertainty principle, declared, "We can know where the next science station is going to be, or we can know when we are going to stop for it, but we can never know both at the same time." Under that, he wrote: "The next station will be at 90 degrees N., so we cannot possibly know when we're going to get there."

From now on, everything took a back seat to getting to the Pole. Charting the rise of the warm Atlantic Layer as it closed in on the ice pack, separating copepods from amphipods, looking for polar bears, tracing the radionuclide path through the halocline, assaying the effect of clouds on climate, remotely sensing the shrinking of the polar ice cap — all these receded to the back of our minds, though most hands remained busy with them. We thought of nothing but the Pole, and since none of us actually knew what the Pole would be like when we got there (except Frank, who had been there on the *Oden*, and Ikaksak, who wasn't saying), we had to find something familiar on which to focus our imaginings. For that, we had Doug and Louise's wedding.

Fiona and Janet spent most of the morning making paper flowers from Kleenex and cutting out cardboard wedding bells and covering them with tinfoil. The wedding and the reception were to take place in the helicopter hangar, and a large contingent of double and triple PhDs was conscripted to hang the bells and bunting, set up the tables, and find a sound system. There was to be dancing. Stefan was entrusted with Doug's camcorder. Janet mysteriously produced an electronic keyboard. The ship's cook was consulted about the menu. The cook began to talk about roast beef and ice sculptures. Cold cuts, the captain told him, cold cuts and potato salad. The galley staff set to peeling the wilted outer leaves from hundreds of heads of iceberg lettuce, while the cook consoled himself with desserts.

Eddy and Knut were busy in the boardroom devising another new science plan. The *Polar Sea*'s propeller had been "secured" — Captain Grandy's word for it — and the ship was now using her gas turbines to maintain headway. We should still, the captain said, "be able to complete our journey within the allotted time."

No one quite knew what that meant, but each of the possibilities was discussed in the hangar as we hung bunting and cut out bells. Did it mean we would still be in Point Barrow on the 21st of September?

"If the *Polar Sea* is using her turbines," said Frank, "then we won't be going to Barrow at all. We certainly won't be going back through 1,700 nautical miles of ice. I suspect we'll be going straight on through to Norway."

To Norway? That meant continuing past the North Pole, once we got there, to come out of the ice in the Greenland Sea. There was a lot less ice on that side, but there were still 600 nautical miles between us and the ice edge. Could the *Polar Sea* make it? And what would it mean to our science plan? We were being carried along by the ice almost as helplessly as Nansen had been. I marveled at how little had changed in a hundred years of Arctic exploration: the threat of ships caught in ice, the uncertainty, the strain of close quarters. At least Nansen had been able to calculate when he might be free of the ice, but not where. We had refined Heisenberg's uncertainty principle even further. We didn't know where we were going after the Pole, and we didn't know when we would get there.

Robert Edwin Peary is still officially listed as the first person to reach the North Pole by foot, and a few people still believe that he made it. Peary claimed to have attained what he called "the goal of the world's desire" on April 6, 1909, after a quarter-century of frustrated ambition. He had traveled to the Pole by dogsled from Ellesmere Island, he told the world, for the last 240 kilometres in the company of two Inuit guides and Matthew Henson, his black "personal assistant." Then he had turned around and walked back. "The determination to reach the Pole," he wrote upon his return, "had become so much a part of my being that, strange as it may seem, I long ago ceased to think of myself save as an instrument for the attainment of that end. To the layman this may seem strange, but an inventor can understand it, or an artist, or anyone who has devoted himself for years upon years to the service of an idea."

Peary's claim was first challenged by Dr. Frederick Cook, who claimed to have been to the Pole himself in 1908. Cook's announcement infuriated Peary, of course, and the latter launched a vicious and successful character assassination campaign against "that G-d-m s-o-b Cook." The controversy boiled away for years: in 1934, one of Peary's supporters, Jeannette Mirsky, wrote *To the North!*, dismissing Cook's claim, deducing from his diary and photographs that he was never out of sight of land. But the book was withdrawn from publication when Cook threatened to sue, and it wasn't released until 1948, after Cook's death, under the title *To the Arctic!* In the end, Cook's account was officially discredited, as was his earlier claim to have been the first person to climb Mount McKinley. This left Peary in sole possession of the Pole.

With Cook out of the way, no one needed to take a close look at Peary's claim. This was fortunate for Peary, because it gave him almost 60 years of undisputed glory. But that glory began to fade as other explorers, following in what they thought were Peary's footsteps, realized that Peary's footsteps weren't, in fact, there.

One of the first modern writers to question Peary's claim was Farley Mowat, whose book *The Polar Passion*, an anthology of various explorers' accounts of their attempts to reach Highest North, appeared in 1967. Mowat didn't even include Peary's account, which he said was nothing but "frauds," but championed Cook, "a lone wolf and an outsider first and last," over Peary, who "had the Establishment on his side" and "was naturally amoral, and therefore icily effective." Ralph Plaisted also called into question the accuracy of Peary's account. Plaisted snowmobiled the last 500 nautical miles to the North Pole "for fun" in 1967, leaving on March 28 and arriving at the Pole on May 4 — making an average of 13.5 miles a day, a remarkable achievement, and one that gave him some insight into Peary's claim to have walked in a straight line 133 miles in four days — an unbelievable 33 miles a day — and then followed his own footprints back to his starting point. Wally Herbert, who, with three companions, made it to the North Pole by dogsled in 1969, was even more vociferous in his attack on Peary. In *The Noose of Laurels* (1989), Herbert examined Peary's published account as well as the private diaries, and concluded that Peary had missed the Pole by 50 nautical miles, but could not "admit first to his five bone-weary companions who had risked their lives to get him to that point, and later to his supporters and public, that 'the greatest Arctic explorer of the Age' ... had made an error of navigation."

It was odd reading Cook's and Peary's descriptions of the North Pole at the North Pole, and to realize that they were fabrications. "Nearing the Pole," wrote Cook, "my imagination quickened." Indeed it did. "We were all lifted to the paradise of winners as we

Robert Peary, "the greatest Arctic explorer of the Age"

stepped over the snows of a destiny for which we had risked life and suffered the tortures of hell. Constantly I watched my instruments in recording this final reach. Nearer and nearer they recorded our approach. At last we touched the mark! We were at the top of the world! The flag is flung to the frigid breezes of the North Pole!"

At midnight, the GPS, which had not crashed, read 89 degrees 52 minutes N: we were 8 miles from the Pole. The sky was a brilliant blue, for once, and the sun glinted off the ice when we looked aft, toward the *Polar Sea* following in our wake. Mike Hemeon put down the handset.

"I just spoke to the captain," he said. "I expect he'll be up shortly. He only put his head down an hour ago."

To our astonishment, we were watching for a third ship on the horizon. Nine days earlier, we had received word from Ottawa that a Russian icebreaker, the *Yamal*, was somewhere in the vicinity of the Pole, and since then we'd been trying to raise her on the radio. So far, she had not responded. We were beginning to feel uneasy about her. Was she in distress, unable to reply? Was she under orders not to make contact with us? Had we been watching too many movies like *The Hunt for Red October*?

"Looks like a puff of smoke over there on the horizon," somebody said.

"Better not be," said Mike. "The *Yamal* is nuclear-powered."

We took up our glasses and scanned the horizon. Every ridge looked like a ship's funnel. Every cloud looked like smoke.

Captain Grandy came up at 12:30, looking tired. He took up his usual position at the starboard end of the bridge.

"What's our latitude?" he asked.

"89-55-5, sir," said Mike.

"Our heading, Les?" he asked the quartermaster at the wheel.

"017, sir."

We came upon a broad strip of broken ice running across our path: the track of another ship, recently passed. The *Yamal*. The track made a huge circle in the ice, like a game of fox-and-geese, with the much-imagined North Pole at its center. Captain Grandy decided to follow it.

"Hard a-starboard."

"Hard a-starboard."

An hour later, we could see two stacks poking above the horizon a few points off our starboard bow. The track we were following curved around toward them. There was no smoke from her stacks at all, of course, which only made her seem more ominous. (I wondered why she even *had* stacks.) The captain spoke into the radio again, calling the *Yamal*, but there was no reply. She was a ghost ship, position without momentum.

Mike, standing at the GPS readout behind the captain, called out: "89-55-662, we're losing latitude fast, Captain. Now it's 55-642."

"20 degrees starboard there, Les."

"20 starboard."

"Hard a-starboard. Work up through that lead on the starboard bow."

"Starboard bow."

"Midships."

"Midships."

"Keep her coming to starboard."

"Coming to starboard."

"What's our reading now, Mike?"

"89-55-659. Coming up again slowly."

"Very good. Continue working to starboard, Les."

"Working to starboard."

"55-652, we're starting to decrease again," said Mike.

"What's the bearing, Les?"

"324."

"55-684, going up again."

"I'll find that son of a bitch," said Captain Grandy, "if it takes all night."

In a way, the best argument that Peary actually may have made it to what he thought was the North Pole is that he denied getting there at all. Unlike Cook, who said his instruments recorded his approach to the Pole until "at last we touched the mark," Peary spent a long time explaining the impossibility of getting his instruments to read 90 degrees. "The last march northward ended at ten o'clock on the forenoon of April 6," he wrote. "After the usual arrangements for going into camp, at approximate local noon, of the Columbia meridian, I made the first observation at our polar camp. It indicated our position as 89°57'."

Peary was 3 nautical miles from the Pole. Did he press on to accomplish his ambition of the past 30 years? No, he set up his tent. "I was actually too exhausted to realize at the moment that my life's purpose had been achieved." In fact it hadn't been achieved, unless Matt Henson, who had arrived there some hours earlier, was right and had actually camped at the North Pole, not 3 miles short of it. If that were the case, though, then Henson would have beaten Peary to the Pole, and Peary would not have that. Peary went to sleep.

When he woke up two hours later, he took a reading, which he doesn't record, then loaded a sleigh with his instruments — a sextant and an artificial horizon — and set out with the two Inuit companions, leaving Henson behind. After a while he stopped and took a second reading: "These observations," he wrote, "indicated that our position was then beyond the Pole." Peary then concluded that he had passed over the Pole. He turned his team around and went back, passing, as he imagined, over the Pole again. He did this 13 times, crisscrossing back and forth over an area of 26 square kilometres within which he thought the Pole was located: "In traversing the ice in these various directions as I had done," he concluded, "I had allowed approximately ten miles for possible errors in my observations, and at some moment during these marches and countermarches, I had passed over or very near the point where north and south and east and west blend into one."

In a footnote, Peary anticipated the ensuing controversy by protesting too much. "No one," he wrote, "except those entirely ignorant of such matters has imagined for a moment that I was able to determine with my instruments the precise position of the Pole, but after having determined its position approximately, then setting an arbitrary allowance of about ten miles for possible errors of the instruments and myself as an observer, and then crossing and recrossing that ten-mile area in various directions, no one except the most ignorant will have any doubt but what, at some time, I had passed close to the precise point, and had, perhaps, actually passed over it."

Close to the precise point? Arbitrary? Perhaps? By Peary's own admission, then, he didn't have the slightest idea whether he was at the North Pole or not. There is no mark. Peary knew that. No instrument in existence then could record the mark that was the North Pole. Perhaps no instrument in existence now could do so, either.

"Midships."

"Midships."

"Bearing?"

"354."

"Bring her up to north."

"Up to north. Whoa!"

"What's the matter, Les?"

"When she got on north," Les said, looking at his GPS, "she flashed all nines at me."

"Well, keep her on the nines, then."

"007 now, sir."

"Put her back on north."

"Aye, sir."

"She's starting to jump up good now," said Mike, forgetting in his excitement to say "sir." "She's at 55-96."

"Yes!"

"89-57," Mike said. "Three miles to the Pole."

Wally Herbert, in *The Noose of Laurels*, pointed out dozens of places in Peary's account that don't stand up to scrutiny. The only part he didn't tear to shreds was Peary's confession that he never actually measured 90 degrees. Could this be because, when Herbert himself arrived at the Pole on April 5, 1969, he couldn't measure 90 degrees either?

In his own account, *Across the Top of the World*, Herbert prepared us, as Peary had, by explaining how difficult it is to calculate position at the Pole (he was using a theodolite rather than a sextant, but the problems were similar): "If your calculation of the longitude is slightly out," he explained, "then the time at which the sun crosses your meridian — in other words, the time at which the sun is due north — is wrong, and so you head in the wrong direction." Lines of meridian are only a metre or so apart when you are close to the Pole, and being out a few degrees of longitude is almost inevitable. The distance between your eyes can be 2 degrees of longitude. Geography deserts you at the Pole. Eventually, Herbert continued, "you increase your errors progressively until you spiral into almost a complete circle. This is what happened to us on this particular day."

Herbert and his companions, 7 nautical miles from the Pole, three weeks behind schedule and having already radioed a message to Queen Elizabeth that they had made it on April 5, suddenly realized they hadn't the slightest hope of finding anything they could confidently call the North Pole. Herbert's own account is virtually identical to Peary's:

> In desperation, we off-loaded the sledges, laid a depot and took on with us only the barest essentials, just enough for one night's camp. . . . With the lighter sledges we made faster progress, and after about three hours estimated that we must surely be at the Pole, possibly even beyond it. So we stopped, set up our tents, and did a final fix which put us at 89°59' N, one mile south of the North Pole on longitude 180. In other words, we'd crossed the Pole about a mile back along our tracks. But the drift was now with us, so we must surely cross the Pole a second time as we drifted overnight. We got into our sleeping-bags and fell asleep.

By 3 in the morning, the sun behind us, the temperature was exactly zero. Between us and the Pole the pressure ridges formed impenetrable sinews of ice nearly 12 metres thick. They seemed to be guarding the Pole, like tendrils of thorns in an Arctic version of Snow White.

The world was ice, a universal congelation, a suspension of the laws of nature. The ship's bow would ride up on a multi-year floe, the floe would not crack, and we would slide back into our own past to try again, like a whale bent on beaching. We were at 89 degrees 58 minutes N.

"Two miles to the Pole," Mike called out.

"How's the longitude?" The lines of longitude were now so close together that if the ship skewed a few feet to one side off a floe, the GPS would race through them wildly.

"Holding pretty good."

"Bearing, Les?"

"357."

"Hold her steady."

"Steady as she goes."

We had all come up to crowd the bridge, 35 of us standing silently along the wooden railing, staring ahead through the glass. Lines of longitude continued to collapse in upon themselves; time zones contracted. We were a mile and a half from the Pole. A mile. Half a mile. We were 800 metres from it. We could all see it, each of us looking at different spots on the ice.

"How far now?" called the captain.

"Half a cable!"

"Goddamn it, the length of the ship!"

"These are hard floes, Captain."

"Play snooker with 'em, Les!"

"We're 50 metres from the Pole, Captain, less than half the ship's length. We just beat the *Oden*!"

The GPS gave 89-59-97, the highest it would ever reach. Our bow rode up on the floe again, paused for a long instant, and slid off to starboard. The captain reversed engines, the GPS raced wildly backward — 89-59-729 — and we charged the floe again, the Pole at its dead center. Again we slid away. The captain shouted: "Once you find it, it's gone!"

"89-59-6! We're losing it, Captain!"

"Damn!"

Eddy had been watching calmly. Now he turned to Captain Grandy and said: "You know, theoretically it's impossible to reach the Pole."

The captain looked at him as though at a madman and said nothing. Mike Hemeon nearly shouted at Eddy: "No it isn't!" And he ran through space and time to the port side of the bridge, behind the line of watchers at the rail, crossing a hundred meridians and passing through seven time zones, and pointed down to a spot on the ice just off our port bow:

"That's it there," he said. "That's where we'll put the North Pole!" And he was as right as anyone has ever been.

The North Pole was our longest station. We would do 12 hours of science, then have 12 hours off for the wedding. Eddy had scheduled something for every program: four CTDs, two rosettes, ice cores and beams, Challenger and Seastar pumps, a box core, zooplankton hauls, snow sampling for albedo and iron. With 4 kilometres between us and the ocean floor, it would take three hours for each box core, and three more hours for each rosette, which couldn't be lowered until the box cores were up, in case the cables became entangled. The captains, Eddy said, were nervous about spending too much time in one place; it was winter, the ice was thickening daily, and, for the *Polar Sea*, there was the question of fuel. Even stationary, the ships burned a vast amount of oil. We began teasing Doug and Louise that their wedding would have to take place while Doug was standing at the starboard winch, raising the rosette.

When the science work was done, Knut brought us up to date on the *Polar Sea*'s propeller problem. "There are five bolts gone from the retaining ring on the starboard shaft," he said, "and they're getting play in the sleeve. Coming down from the Lomonosov Ridge on Saturday, the prop didn't seem to want to pitch properly: there was a lot of vibration, and so they decided to feather it — just let it spin on its own with no power from the engines, with the ice feeding through it. Then at 0230 on Sunday morning there was a whopping bang, followed by even more intense vibration."

The night before, when the *Sea*'s diving team had gone down to see what had caused the bang, they found that one of the blades from the starboard propeller had flown completely off, hitting the center prop as it flew and damaging two of that prop's blades. That would account for the bang, all right.

"Not only that," Knut continued, "there also appears to be a problem with the retainer ring on the port screw, the same damage that first showed up on the starboard prop. Half its retaining ring is gone, and the rubber lining has slipped out and is shipping sea water. As you can imagine, this is not good news."

Shipping sea water had an ominous ring to it, a kind of historical resonance. We could imagine the *Titanic*'s captain cabling his last message to the world: "Shipping sea water!"

"What it means, essentially," said Eddy, "is that we're abandoning the plan to go back through the ice to Point Barrow. We're going to continue straight on across the Pole, to exit the Arctic Ocean via Fram Strait. While we have no plan to alter the duration of the voyage, this too could change. The track of the *Sea* has been drastically altered, and it is our obligation to see her safely out of the ice."

When the meeting broke up, most of us sat around for a few moments, too stunned to move. Not going back to Point Barrow meant that half the science programs had just been canceled. This was an enormous blow. But if the duration of the expedition had not changed, the scientists would have to come up with new plans for the eastern half of the Arctic Ocean instead of the unknown western portion. Rather than tapping into totally new sources of information, they would be sampling areas that had been visited by other scientists off and on for more than 50 years.

The helicopters had been wheeled onto the flight deck, out of the way. The hangar was strung with bunches of paper flowers, cardboard wedding bells wrapped in tinfoil, and strings of crepe paper. The head table, hung with bunting, had been placed along the starboard bulkhead, covered with a white tablecloth and set with places for half a dozen people: hand-printed cards placed Doug and Louise in the middle, the two captains on either side of them, then Eddy and Fiona (Louise's maid of honor). Along the opposite bulkhead, a similar but longer table held the buffet: tray after tray of cold cuts and salads, breads and desserts. The roast beef cold cuts were deliciously pink. The lettuce looked fresh. As Janet played the "Wedding March" on the electronic keyboard, Captain Grandy moved to the small lectern, the rest of us crowded around, and then Doug and Louise, nervous but smiling, walked arm in arm up the middle of the hangar and stood before the captain. Doug was wearing a black tuxedo with a white tissue-paper boutonnière in his lapel. Louise wore a black satin jacket and white blouse, with Victorian lace at her throat and wrists. Eddy and Fiona took their places beside them. Captain Grandy, in exercise of the power vested in him by the government of the Northwest Territories of Canada, cleared his throat and plunged in:

"I, Douglas Gordon Sieberg. . . ."

"I, Douglas Gordon Sieberg. . . ."

"Take you, Louise Margaret Adamson. . . ."

"Take you, Louise Margaret Adamson. . . ."

What followed the wedding is difficult to explain. Some say the Russian ship, the *Yamal*, sent one of its helicopters over to buzz us; others claim that we sent one of ours over to buzz them. Whatever the case, the *Yamal*'s commander, Captain Smirnoff, finally answered one of Captain Grandy's calls. It was announced that the *Yamal* was 18 nautical miles away and we were going to join her for a trilateral party on the ice, the first time in history that icebreakers from three nations would tie up together at the North Pole.

"I thought we were *at* the North Pole," said Steve Hemphill after the captain's announcement.

"Maybe we've drifted off during the night," I ventured. "Where was the *Yamal* when you buzzed her?"

"She'd moved off the Pole to let us have it to ourselves," Steve said.

"Maybe she's drifted back. What does she look like?"

"Huge," said Steve. "Twelve decks. Big and black and red, with a big white shark's grin painted on her hull."

"Go on."

"I swear," said Steve. "And there was a huge apple on the ice beside it, almost as high as the bridge, and there were kids dancing around it, holding hands."

"Uh-huh."

"And then the apple opened up in sections, like it was sliced, and inside it was a huge stage with more kids dancing and playing electric guitars."

He looked at me. "No, really," he said.

A few hours later, we were anchored to an ice floe beside the biggest ship I had ever seen: a huge, black-and-red, top-heavy icebreaker, 12 full decks above the waterline, and with a big white toothy grin painted on her bow. She looked as though she could swallow ice. Steve's gigantic apple was gone, but his description of the *Yamal* had been so accurate that I did not doubt the existence of the apple. There were still two large red speakers on tripods set up on the ice beside the ship, and a row of smoking barbecues (could these be what we had seen on the horizon?) with lines of fidgety children waiting for reindeer meat and Coca-Cola. The *Yamal* had brought 55 performing arts students from Russia, with their families, to the North Pole in celebration of International Family Week. The apple had been built on the ice as a kind of stage, and the performers had given a show that had been televised back to Russia via satellite.

Seven of us were in the company of a large man named Leonid, a television producer from the Ostankino TV Center, in Moscow, who had walked over to the *Louis*, had a tour, and now wanted to show us his television studio on the *Yamal*. He spoke a heavily accented but reasonably fluent English. We threaded our way through the children, who were grouped around the barbecues, and made our way to the foot of the *Yamal*'s gangway.

"They are music and dance students," Leonid said. "From all over Russia. We are making a television production here. The apple is the world. It is very interesting, very difficult."

Western disco music blared from the speakers, and knots of Russian sailors and American Coast Guard crew members stood tapping their boots on the ice and drinking brandy toddies, euphemistically called "tea," from a row of steaming samovars perched on a table on the ice. Between the barbecues and the samovars, children in brightly colored ski outfits discoed with each other and their parents. The language barrier had long since been overtaken by the sound barrier. We were stopped at the gangway by the *Yamal's* quartermaster, a small, bristle-haired man wearing dark glasses, who eyed Stefan's video camera either suspiciously or covetously, we couldn't tell which. Leonid listened to him and then turned to us.

"You are supposed to wait for a group tour," he said, "but I told him you are important television people from America. Let us go up, please."

The quartermaster did not seem to question that a Russian nuclear ship, tied to an ice floe at the North Pole, would be visited by important television people from America. At the top of the gangway we entered the ship through a kind of lobby, like the entrance to one of the Russian-designed hotels I had seen in northern China. It had the same sweeping staircase, only smaller. Under the stairway, where the check-in desk ought to have been, there was a small tuckshop selling *Yamal* sweatshirts, *Yamal* coffee mugs, and polar bear lapel pins. It was closed. Leonid began to lead us to the stairs — "First to cabin, drink vodka," he boomed — but we were diverted from this pleasant prospect by an officious-looking Russian officer in a military uniform buttoned to the throat. He said something to Leonid and then consulted his clipboard.

"Please," he said, "you will all wait here for translator, who will take you on exposition."

Leonid began to remonstrate with the officer — maybe we were now representatives of a big American vodka company — and while they were discussing the situation a large, carefully made-up, recently blonde woman appeared wearing pink track pants and a white sweatshirt that said, in English: "Polar Expedition 1994 — Yamal." She told us she was the translator and exposition guide: her name was Irena, and during the school year she taught English in Murmansk.

"If you don't mind, let us begin our tour."

We left Leonid waving his arms at the officer as Irena took us along a corridor and down a nearly vertical ladder to one of the

Yamal's vast and crowded engine rooms. The noise was deafening, but her voice rose above it: "Is steam generators," she shouted. "All the steam is here generated on ship. Come."

We followed her along another corridor into a second room where the machinery was larger and the noise louder. "Is steam turbines," she screamed, "to drive the propellers."

I turned to Frank. "The *Yamal* is a steamship?" I mouthed. Frank shrugged. He seemed unsurprised that the most modern icebreaker in the world had traveled into the Arctic propelled by the same technology that had driven the *Jeannette* and, for that matter, the *Robert E. Lee.*

The third room was lower and larger, with monolithic machinery and noise so loud it hurt to clench our teeth. "This is electric generators," Irena bellowed. "All electricity on ship is generated by these machines." The generators, each about the size of a tractor, were old, probably built in the 1950s: they had obviously been lifted out of another ship and integrated into the *Yamal's* system. "Now if you don't mind, let us continue our tour."

"How many rooms on the *Yamal?*" I asked Irena, feigning lack of interest.

"One thousand," she said.

It was hard to imagine a level lower than the one we were on, but after a few twists in the corridor a trap door appeared with a ladder leading down into a cavernous engine room. Here the noise was so loud that we clasped our hands to our ears, then checked for blood. The roar was demonic. It seemed to rise and fall in waves, but that could have been our own hearts pounding in our ears. We looked painfully at Irena and tried to lip-read her words. I caught "engines" and "75,000 horse-charge," and then gratefully followed her back up the ladder.

In the corridor she stopped and grouped us. "How many would like to look into nuclear reactor?" she asked. We all muttered our enthusiasm. "Until this year," she said, "wisitors could go right into reactor room to look. But now insurance is wery high, so you must to look through glass window. This way, please."

She led us into a long, narrow, white-painted room. Pipes and wires had been duct-taped to the curved bulkheads over our heads. We were evidently in some kind of enclosed catwalk that overlooked a central cathedral-like chamber, sensed rather than seen on the other side of the bulkhead. Halfway down the room was a small glass window, beside which stood a smiling technician wearing blue coveralls and holding a digital Geiger counter. He mimed turning on the Geiger counter, held it to the window, and then motioned to one of us — Peter — to look at the readout. "Zero," Peter said, beaming back at us.

We took our turns peering through the glass, down (could there still be more down?) into an enormous white chasm filled with green tubes and wires spiraling down the walls like vines in some overexposed mechanical garden. In mid-air the vines split into two coils; each led to one of two huge hot-water tanks anchored to the floor. Irena leaned over my shoulder and, pointing to the hot-water tank on the left, said, "One reactor," and then pointing to the one on the right, said, "Two reactor." Objects A and B. It occurred to me that I was looking into what had been one of the best-kept secrets in the world. Spies had risked their lives for this view. Captain Brigham, though he had spent weeks aboard Soviet icebreakers, had not been permitted this view. And yet here we were, important American television people, being escorted through the ship's entire below-decks domain, nuclear reactors and all. Didn't they know we had cameras? Irena even ushered us into the computerized control room — an area as big as the *Louis*'s dining hall lined with digitized schematics of the *Yamal*'s wiring system (in yellow), propulsion layout (in red), heeling and fresh-water tanks (in blue), and other diagrammed and backlit perplexities. One readout caught my attention: "Reactor temperature: 13,000 degrees C."

"How much fuel do you burn in one day?" Irena asked Frank, interpreting for one of the engineers.

"About 100 cubic metres," said Frank.

"We about 100 grams," Irena shot back on her own.

Back in the corridor, we ran into Leonid and his friend Igor, who spoke poorer English than Leonid, but more of it. There was no more talk of vodka. Igor made a wide, sweeping square with his arms and said, "Russian steel. Very big." We all nodded. Then he said, "Builded in Leningrad two years ago. Was then called *October Revolution*. How long you take to go here?"

Again Frank replied: "About a month."

"We leave Murmansk for North Pole," said Igor, holding up four fingers and bending one of them in half: "These many days."

"We came the hard way," said Frank.

After the *Yamal* tour I returned to the *Louis* in a vacant mood. It was 8 P.M., and the sun was at its lowest point, barely clearing the horizon. Clouds infused the sky with a false but strangely comforting sense of dusk. When I heard my name being called, I turned and saw Captain Brigham walking behind me.

"It's amazing, isn't it?" he said when we were walking together. Normally he wore a *Polar Sea* baseball cap with scrambled egg on the peak, signifying his rank; now he was wearing his real uniform cap, and I could see his dress uniform under his down-filled parka. "Think of it," he said. "Think of the historical significance of it. The top ice-breakers of the three biggest polar nations on Earth, tied up to the same floe at the North Pole. My God," he said, shaking his head. "If anyone had tried to organize this, it would never have happened."

We parted company at the *Louis*'s gangway, and I watched him walk on alone toward his own damaged ship.

In the *Louis*'s lounge, Captain Grandy was entertaining Captain Smirnoff and three of the Russian officers. Most of the scientists and Coast Guard officers were gathered around their table, drinking toasts to international friendship and cooperation and listening to Captain Smirnoff, who spoke through his official translator.

"I have been to the North Pole seven times," he said. He showed Captain Grandy a list of the ships that had been to the Pole; most of them were Russian ships. The *Arktica* was the first. The *Yamal* was listed as numbers 13, 14, and 15; the *Louis* was number 16, and the *Polar Sea* was number 17.

"We have several people on board who have been here before," said Captain Grandy. "Dr. Jones and Dr. Zemlyak were here in 1991." Peter and Frank stood forward and smiled. "And Ikaksak, here, has been to the North Pole five times now."

Ikaksak stood up uncertainly and looked at Captain Smirnoff.

"Where do you live?" asked the Russian captain.

"Resolute Bay," Ikaksak replied.

"What is it like to live so far north?"

Ikaksak smiled. "Very hard," he said.

16

The Channels

The ice cracked behind us and was driven with force towards the
north; a breeze sprang from the west, and . . . a passage towards
the south became perfectly free.

— MARY SHELLEY, *Frankenstein*

WE LEFT the North Pole on August 24, after breakfast. Even without considering the *Polar Sea's*
weakened condition, the best plan was for the
three ships to go together, with the *Yamal* in the lead and the two
smaller ships following in the wide path she made through the ice. It
wasn't exactly an escort, but the *Yamal* was going a lot more slowly
than she could have, and the *Polar Sea* couldn't have made it out on
her own. I suppose we could have waited until the *Yamal* was out of
sight and then followed her track, but that would have felt like skulk-
ing, and we were too elated to skulk. The *Louis* could have stayed behind to continue doing science in high latitudes, but it wouldn't have
been a good idea for us to be alone in the ice either, especially with
winter setting in. The temperature was −2.8 degrees C and falling.

The question of the day was, What would the *Louis* do when the
Polar Sea was safely out of the ice? The *Yamal* would continue along
the 35th meridian, which would take her directly to Murmansk; the
Polar Sea would veer west and make a beeline either to Tromsø or to
Reykjavík, for refueling. From there she would return to Seattle for
repairs, either through the Northwest Passage and Bering Strait, or by

continuing down through the Atlantic to the Panama Canal and then up the West Coast. No one seriously thought she would tackle the Northwest Passage in September with a broken propeller, but her officers kept up a running discussion about it.

The *Louis* could either stay with the *Polar Sea* all the way to Iceland or Norway and then return home to Halifax, or else bid her farewell at the ice edge and then turn back to conduct more science stations in the open water north of Spitsbergen, or even go back into the ice to do some work north of Greenland. We had plenty of fuel, and we weren't expected back until the end of September. Captain Grandy began seeking permission from the Norwegian government to work in its territorial waters north of Svalbard. If permission didn't come, we would still have to decide whether to call it quits and go home with the *Sea* or else go north into the ice again to continue the expedition.

"There's work to do in the Laptev Sea," said Eddy at the science meeting that day. "There's work to do in the Greenland Sea. Heck, there's work to do everywhere. We still have four weeks, lots of fuel, and one of the finest oceanographic teams ever assembled."

Meanwhile we continued edging our way south. It was odd to have the sun ahead of us, off the starboard bow, obscured though it was by a white veil of snow. I found myself drawn more to the forward deck now. The thin snow was blowing horizontally, as it often blew back home; little Ls of it were accumulating in the window wells and along the decks. And the kittiwakes had returned, or else had been replaced by their eastern cousins, to continue their wheeling and diving, exactly like ring-billed gulls dipping for alewives on Lake Ontario. I realized I was homesick. I didn't want to go back into the ice. It had been all right when we were in it, when there had been no talk of going home, when we were heading north. But now that the possibility of an early return had been raised, going home was all I could think of.

"You've got a bad case of the Channels," said Steve Hemphill, who, like me, was drawn more and more often to the foredeck. No ice problems and no science stations meant there wasn't much for him to do on the flight deck, either.

"The Channels?" I asked.

"Channel fever," he said. "Once you start thinking about home, you just want to get there. You don't want any delays."

"It's going with the flow," I said. "Heading north through Bering Strait, we were going with the current and couldn't wait to get to the Pole. Now, in Fram Strait, the current is flowing south and it feels like the water is hurrying us home."

"The captain feels it too," said Steve. "He can smell Halifax."

The next day, Steve flew me over to spend a few last days on the *Polar Sea*. I had barely stepped out of the helicopter when the fire alarm started ringing and a voice came over the pipes: "This is not a drill. Proceed to your fire stations immediately. This is not a drill."

I had no idea where my fire station was. I'd been on the *Louis* for my fire drill. I joined the flight deck crew, who were streaming forward to the crew's lounge, where I picked up a group of scientists who seemed to be heading up to one of the upper decks. There was an air of solemnity on the ship, not alarm, but concentration. The fire klaxon was particularly loud in the stairwells, and no one spoke. After climbing a few levels, we found ourselves outside the officers' lounge, where we decided to wait out the emergency. Inside the lounge, several officers were watching *Home Alone 2* on television, and we sat down and watched with them in an atmosphere of surreal calm. After a few minutes, Mike Powers came in and announced that the source of the problem had been located and the fire was under control.

"It was the brake on the starboard propeller," he said when he had our attention. "We've been going fast, about 6 knots, trying to keep up with the *Yamal*, and the water has forced the prop to turn even though the brake was on. That caused the brake lining to smoke, the fumes exited the ship in the normal fashion, but then re-entered through the ventilation system, and the smoke alarm went off."

We cruised south along the 35th meridian for the next 40 hours and then stopped for a science station. We were down to nearly 86 degrees N, still in the ice but only a few nautical miles from the ice edge — four weeks to get in, three days to get out. Following the *Yamal* had been like driving a vw beetle behind an 18-wheeler. But this is where we would part company. The huge Russian ship cut a wide circle in the ice, like a truck turning in a parking lot, gave us a blast on her horn, and continued on to Murmansk. Her big grin seemed to widen. I watched her go with mixed emotions. I would have liked to see Murmansk.

I would have liked to see Tromsø again, too, that quaint European city nestled in a scattering of islands at the northern tip of Norway. Maybe I could send a message to my wife to ask her to join me. I had spent a week in Tromsø in May 1989, waiting for the *Polarstern*, and had been seduced by its balmy climate so far north. It had cobbled streets and trellised houses with green lawns and deciduous shrubbery. Its downtown had modern banks and department stores, little shops and large hotels, and the world's northernmost brewery: the label on a bottle of Mack-Øl said, in English, "The First Beer to the North Pole," although I never learned who had taken it there. There was a statue of Roald Amundsen in the town square, complete with pigeons. His mustache had been larger in Nome, but in Tromsø he was not encased in granite. Tromsø was on the same latitude as Tuktoyaktuk, but thanks to the Gulf Stream its climate was more like that of central Ontario. Global warming might change that.

It was another long science stop, nearly 12 hours; the rumor on the *Polar Sea* was that the *Louis* had requested an extra five hours despite the *Sea*'s hurry to get out of the ice and back to Seattle for repairs. Hearing the rumor made me feel uneasy. Whether or not it was true, it indicated the tension that had built up between the two ships. Later, when I heard the truth, I felt worse. The *Louis*'s rosette had malfunctioned, and since we were over the Barents Abyssal Plain (where Knut suspected the warm Atlantic water was entering the Arctic Ocean), Knut had asked Art to tell the *Sea*'s rosette team to do two casts instead of one. Art had simply ignored the request, and when Knut called over to find out how the station was going, he learned that no second rosette cast had been carried out. He ordered it done then, and since the Barents Abyssal Plain is nearly 4,000 metres deep, the extra cast had taken a long time.

I found Art in the crew's mess, drinking lemonade and reading a travel brochure from Iceland. "They're going to let the scientists off in

Iceland," he said when I'd joined him. "At least, that's the latest rumor. Most of us will fly home from there directly, but I thought I'd spend a few days doing geology. I'll put a call through to the university in Reykjavík. If there isn't someone there who can take me around, at least they'll have a decent field guide in the bookstore."

I thought of an article on volcanic islands by John McPhee that I'd read in the *New Yorker* a few months before our trip. "Iceland must be pretty interesting from a geological point of view," I said.

Art nodded glumly. "Actually," he said, "all I really want to do is get off this goddamn ship and on with my life."

Terry and Erk came into the mess after Art had left. Terry had solved the problem of the different-colored melt ponds. "I was talking to Michel Gosselin this morning," he said. "He told me about ice algae. There is a species that is golden yellow. When this golden yellow species inhabits multi-year ice, which is blue, it turns the ice green: yellow and blue make green." He was amazed by the simple beauty of it. Science doesn't get much purer than that. "Michel is trying to work out why this yellow species grows on some ice and not on others; it probably has something to do with the amount of light that penetrates through the ice." He said he loved this mixing of disciplines. "It's fabulous when a biologist can teach a geologist something about ice."

Yesterday, through binoculars, Erk had seen two large pieces of wood sticking out of the ice and had gone out in one of the helicopters to collect them. They were about a metre long and half a metre in diameter, clearly pieces of pulpwood that had been cast up on some shore or other for a long time before becoming frozen into the ice and carried out to sea. They were worn smooth and bleached white, jagged at both ends and studded with gravel.

"They are too big to be taiga," said Erk, "so they must have come from south of the tundra. When I get a chance to look at their ring structures, I might be able to determine where they grew."

Tree ring patterns are distinct for different areas of the North, varying with local weather and soil conditions. Trees do most of their growing in June and July; favorable weather during those two months produces wide tree rings, poor weather means tight, narrow rings. "By checking historical weather records, I can distinguish between trees grown in the Yukon and those grown in Siberia." Given the kind of ice in which he had found the wood — heavy multi-year ice that had traveled across the Arctic Ocean — Erk guessed that the wood had come from somewhere along Siberia's Lena River. "The Lena is a long river that could easily carry logs of this diameter into the Arctic Ocean, and there is a lot of logging in the interior of Siberia; I have seen piles of such logs stacked up on the gravelly banks at the mouth of the Lena. It was probably carried into the Laptev Sea, floated into the Laptev polynya, became fast in the ice there, and caught up in the transpolar drift to pass over or near the Pole, and arrive where we found it yesterday. I'd say the journey took about five years."

Now that we were in the northern Greenland Sea, there was nothing the physical oceanographers would have liked better than Norwegian permission to run a line of science stations across Fram Strait at 83 or 84 degrees, north of the area where Arctic Ocean water met Atlantic water, where Atlantic Deep Water was supposed to be forming but somehow wasn't. Brenda and Kathy could also do contaminant work in the Barents Sea, tracing the route taken by cesium 137 from England's Sellafields nuclear plant to the Arctic Ocean. And little was known about nutrient and trophic levels in Fram Strait. There was definitely work to be done, but it would take three weeks to do it properly. Did we have three weeks?

The *Louis* waited in the ice north of Spitsbergen while the *Polar Sea* worked on problems with her rudder. Despite the Channels, I had returned to the *Louis*, ready to spend more time in the ice after parting company with the *Polar Sea*. But permission from Norway never came. On August 30, orders arrived from Ottawa "to continue in company with the *Polar Sea* to Iceland," and then go straight on to Halifax. Knut ordered a final science station, at 75 degrees N, 6 degrees W, while the *Sea* continued south at half-speed; then we would begin tying everything down in preparation for open-water transit and catch up with her.

"Our motto for the expedition," said Eddy at the science meeting, "has always been, 'In together, out together.' We have accomplished an enormous amount during this trip, we have enough data to keep us busy for the next ten years. It's time to go home."

We had indeed collected enough data to keep everyone's computers busy for the next decade, and what we had learned already was of enormous importance. We now knew, for example, that the Arctic was not a desert. It may have appeared lifeless to previous investigators, who, attuned to southern oceanographic models, judged an ocean's productivity by the amount of phytoplankton it contained, and the number of copepods they could pull up in their fine-mesh nets. True, not enough light penetrated the ice to make phytoplankton bloom, but it did not therefore follow that nothing bloomed in the Arctic. Organisms had adapted to the available light under and even within the ice — just as organisms adapt the world over. We had found and identified them. In fact, we had found and identified so many of them that it was no longer possible to argue that there was not enough food in the Arctic to support a fulltime icebound polar bear population.

We knew that contaminants dumped into the oceans in the south eventually find their way into the Arctic, and that increased global warming was pushing more kinds of them farther north than ever. This had given us a better sense of the interrelatedness of the world oceans. The Arctic was not an isolated pool of water that had little to do with the forces that moved heat around the world and controlled its climate. The Arctic Ocean was vitally linked to those forces.

We also had determined that, for reasons not yet understood, much greater amounts of warm Atlantic water were being pushed into the Arctic through the Barents Sea than had previously been the case, and that this warm water was sloshing around in the Eurasian Basin, slopping over the submarine ridges into the Canadian Basin, changing temperature-salinity curves throughout the entire Arctic Ocean, and rising fairly rapidly toward the surface. We knew that if that warm layer of Atlantic water rose high enough to come into contact with the ice, it would change the nature of life on this planet. The climate change models, using no actual data from the Arctic Ocean, predict that the ice cap will be reduced if we get global warming, but their predictions are all based on the albedo effect: as the air temperature rises, they assume, the ice will shrink, less solar radiation will be reflected back into space, and the air temperature will rise some more. That may in fact happen, but we discovered that the ice will also be attacked from below by a rising layer of abnormally warm water: the ice is going to lessen in volume as well as in area, maybe to the point of disappearing altogether, at least seasonally. That is something to worry about.

"What all of that emphasizes," said Eddy, "is the fact that we can no longer think of the Arctic Ocean as a steady state, the same year after year, a one-dimensional entry in a climate model. That idea, I'm glad to say, has finally bitten the dust. I think we've learned to start asking the right questions about the Arctic Ocean. How are a lot of seemingly unrelated phenomena really related? For instance, there is some evidence that waves in the North Atlantic are becoming higher because of the increased number of cyclones in the south, pushing more warm water north, and maybe that's why more warm water is getting into the Arctic Ocean. The Barents Sea is warming up. But while the Barents Sea is getting warmer and has lots of cod, the Labrador Sea is cooling down and has no cod. Why is the Labrador Sea cooling? Is the warm Atlantic water that is getting into the Arctic pushing more cold water out through Nares Strait? We begin to suspect how all these things may be related. We have introduced the notion of variability into studies of the Arctic Ocean, and have thereby shown the Arctic Ocean to be more intimately linked with the Pacific and Atlantic Oceans than anyone has guessed, and we've shown that they are all, in some way, undergoing these perturbations together. That is an enormous stride forward."

Knut, too, felt that much had been accomplished and that most of the work was still ahead. The current debate on global warming, he said, hinged on whether the current warming trend was part of a long-term global warming period, and whether the change was caused by human activity. He hesitated to answer yes to either of them. "We cannot yet say with certainty that the global surface temperature is increasing long-term," he said, "and we certainly cannot yet say that there is a measurable anthropogenic thermal effect."

But despite his scientist's reluctance to make definitive statements until all the information has been assessed, Knut issued one of the most dramatic warnings so far connected with the global warming question. "Given our present understanding of the global heat budget," he said, "which is considerable, there are good reasons to believe that we are entering a regime warmer than any the earth has experienced over the past several million years. In particular, increases in the global mean temperature of several Celsius degrees over the next 50 to 100 years seem quite plausible. Such changes would likely be accompanied by significant changes in precipitation. The net result may well be shifts in population and wealth capable of perturbing the present political order."

That night I woke at 1:30 in the morning, while the ship was steaming steadily south. Unable to sleep, I remembered that I hadn't tied my computer to my desk in the engineers' office. Now that we were in open water, everything movable had to be secured with bungee cords or else it might end up on the deck. The ship was rolling but oddly quiet, and I couldn't figure out why. As I passed the officers' mess, I saw the ship's steward setting up the tables for breakfast; he had raised the sides and placed special nonslip material on them, which meant we were in for a cold breakfast and heavy seas. I poured some coffee and headed up to my office.

On the way, I met Malcolm and Oolatita coming down from the after lounge, where they had been watching movies all night. They'd just finished watching *The Hunt for Red October* again. It was Oolat's favorite movie. The last scene, I remembered, was Sean Connery quoting lines from Christopher Columbus's journal:

> The sea shall give new hope,
> And sleep bring dreams of home.

Instead of continuing to my office, I went out on deck. When I opened the heavy wooden door, I realized why the ship had seemed

so quiet. It was dark outside. There were stars. I hadn't seen real darkness for a month and a half. I caught myself wondering how the quartermaster on the bridge could possibly see in the dark to steer the
ship. Shouldn't we have some kind of headlights?

But when I looked up and saw the Big Dipper, the Great Bear,
Arktos, I realized we had better things to steer by than lights.

Halifax, as Hugh MacLennan observed, owes its very existence to ice.
"The Great Glacier," he wrote in *Barometer Rising*, meaning the
glacier that covered North America during the last ice age, "had once
packed, scraped and riven this whole land; it had gouged out the harbour and left as a legacy three drumlins." The drumlins are now
called Citadel Hill, with its star-shaped fortifications, which we could
see as we edged our way into the outer arm, and the two islands in the
harbor itself, the giant McNab's Island farther out and the inverted
porridge bowl that is George's Island close to the city. Both were alive
with gulls — great black-backed gulls, herring gulls, kittiwakes, old
friends now, reeling in salutation as we threaded our way up the
Stream. Ahead of us, a bright red Coast Guard firefighting vessel
spewed twin jets of water to either side, the traditional naval welcome
that made the tiny ship look like a fire hydrant that had popped its
caps. We stood on the *Louis*'s deck and watched the rainbows that
formed miraculously in her spray, feeling like royalty.

It was September 9, one of those cool, cloudless days that now re-
mind me of Nome in July. We sidled up to the government dock in
Dartmouth; harbor seals squirted out of the way as the gap closed between us and the rest of the world. It was a quiet homecoming: the
dock was sprinkled with friends and families of the crew, and the wel-
coming ceremonies were mercifully brief. Eddy mentioned that
these were not really closing ceremonies at all because, in many re-
spects, the work had just begun. Knut agreed with him. "Our voyage
is not over," he said. "We have undertaken a voyage of discovery; we
have moved from surprise and delight to understanding and delight.
It was an ambitious undertaking. What we did together we did wisely
and well, and we have been greatly blessed in the doing of it."

Malcolm and I took a taxi almost before the ceremony was completed. It was late Friday afternoon, and we were worried that we wouldn't be able to book a flight home before Monday if we didn't make indecent haste to a travel agent. We were driven through the dingier parts of Dartmouth, crossed the bridge that spans the harbor, and then drove through the dingier parts of Halifax before reaching the revitalized center, where the travel agents were. I followed Malcolm's lead: his work with polar bears had taken him to the remoter parts of the globe, and he had accumulated enough air miles for a free flight to the moon. He used the bonus tickets only for research flights, he assured me, but he did know where the best chances for a last-minute flight might be lurking.

We stopped in front of a small, drab building on a sidestreet below the Citadel. The sign on the window gave the name of the agent. We paid the driver, wrestled our kits from the trunk of his taxi, and staggered with them up a narrow flight of stairs so worn and uneven that we kept banging our shoulders against the wall, which after 53 days at sea we hardly noticed. When we entered the little room at the top there were two agents busily feeding paper into computers, which we took as a good omen.

A small, sweatered woman with reading glasses and an oval brooch encrusted with what looked to me like crushed multi-year ice showed not the least hesitation when I told her I wanted to exchange tickets from Point Barrow, Alaska, to Seattle, Washington, to Vancouver, British Columbia, for a new ticket from Halifax to Toronto, for that very evening, if possible, nonstop.

"Aisle or window?" she said.

At the airport, we took turns watching the luggage carts while each of us phoned home. My wife was miraculously, wonderfully in. She had heard of our arrival on the news. We spent a certain amount of time catching up. The baseball strike had ruined the city, she said, and she had passed most of the summer quietly at our cabin in the woods.

"How was the weather?" I asked.

"Rain," she said. "Lots and lots of rain."

Appendix

Main scientific programs and scientists aboard the icebreakers CCGS *Louis S. St. Laurent* and the USCGC *Polar Sea*, during July and August 1994

1. **Atmospheric Chemistry and Physics**
   Peter Brickell, Atmospheric Environment Service, Downsview, Ont.

2. **Biology of the Arctic Ocean**
   Pat Wheeler, Oregon State University, Corvallis, Oreg.
   Kent Berger-North, Axys, Sidney, B.C.
   Lisa Clough, East Carolina University, Greenville, N.C.
   Michel Gosselin, Université du Québec at Rimouski, Que.
   Bonnie Mace, Lamont-Doherty Earth Observatory, Palisades, N.Y.
   Mary O'Brien, Institute of Ocean Sciences, Sidney, B.C.
   Evelyn Sherr, Oregon State University, Corvallis, Oreg.
   Natalie Simard, Université du Québec at Rimouski, Que.
   Delphine Thibault, Bedford Institute of Oceanography, Dartmouth, N.S.

3. **Carbon Dioxide and Climate**
   Peter Jones, Bedford Institute of Oceanography, Dartmouth, N.S.

4. **Climate and Atmospheric Radiation**
   Dan Lubin, California Space Institute, La Jolla, Calif.

5. **Cloud Condensation Nucleus Measurements**
   John Grovenstein, North Carolina State University, Raleigh, N.C.

6. **Contaminant Measurements**
   Robie Macdonald, Institute of Ocean Sciences, Sidney, B.C.
   Eddy Carmack, Institute of Ocean Sciences, Sidney, B.C.
   Louise Adamson, Institute of Ocean Sciences, Sidney, B.C.
   Janet Barwell-Clarke, Institute of Ocean Sciences, Sidney, B.C.
   Liisa Jantunen, Atmospheric Environment Service, Downsview,
       Ont.
   Fiona McLaughlin, Institute of Ocean Sciences, Sidney, B.C.
   Dave Paton, Institute of Ocean Sciences, Sidney, B.C.
   Darren Tuele, Institute of Ocean Sciences, Sidney, B.C.
   Rick Pearson, Institute of Ocean Sciences, Sidney, B.C.
   Frank Zemlyak, Bedford Institute of Oceanography, Dartmouth,
       N.S.

7. **CTD/Hydrographic Section of the Arctic Ocean**
   Jim Swift, Scripps Institution of Oceanography, La Jolla, Calif.
   Knut Aagaard, University of Washington, Seattle, Wash.
   James Elliott, Bedford Institute of Oceanography, Dartmouth,
       N.S.
   David Muus, Scripps Institution of Oceanography, La Jolla, Calif.
   Ron Perkin, Institute of Ocean Sciences, Sidney, B.C.

8. **Marine Geology**
   Art Grantz, U.S. Geological Survey, Menlo Park, Calif.
   Pat Hart, U.S. Geological Survey, Menlo Park, Calif.
   Steve May, U.S. Geological Survey, Menlo Park, Calif.
   Walt Olson, U.S. Geological Survey, Menlo Park, Calif.
   Elizabeth Osborne, Woods Hole Oceanographic Institution,
       Woods Hole, Mass.
   Kevin O'Toole, U.S. Geological Survey, Menlo Park, Calif.

Fred Payne, U.S. Geological Survey, Menlo Park, Calif.
Larry Phillips, U.S. Geological Survey, Menlo Park, Calif.
Bill Robinson, U.S. Geological Survey, Menlo Park, Calif.

9. **Halomethanes in Water, Snow, and Ice**
   Charles Geen, Bovar-Concord Environmental, Toronto, Ont.

10. **Marine Mammals**
    Malcolm Ramsay, University of Saskatchewan, Saskatoon, Sask.
    Sean Farley, Washington State University, Pullman, Wash.

11. **Oxygen Consumption, Denitrification, and Carbon Oxidation Rates**
    John Christensen, Bigelow Laboratory for Ocean Sciences, West-Boothbay Harbor, Maine

12. **Physical Oceanography**
    Eddy Carmack, Institute of Ocean Sciences, Sidney, B.C.
    Fiona McLaughlin, Institute of Ocean Sciences, Sidney, B.C.
    Ikaksak Amagoalik, Institute of Ocean Sciences, Sidney, B.C.
    Oolatita Iqaluk, Institute of Ocean Sciences, Sidney, B.C.
    Mike Hingston, Bedford Institute of Oceanography, Dartmouth, N.S.
    Jim Schmitt, Scripps Institution of Oceanography, La Jolla, Calif.
    Doug Sieberg, Institute of Ocean Sciences, Sidney, B.C.
    Bob Williams, Scripps Institution of Oceanography, La Jolla, Calif.

13. **Radionuclide Measurements**
    Brenda Ekwurzel, Lamont-Doherty Earth Observatory, Palisades, N.Y.
    Kathy Ellis, Bedford Institute of Oceanography, Dartmouth, N.S.
    Rick Nelson, Bedford Institute of Oceanography, Dartmouth, N.S.

14. **Remote Sensing of Sea Ice**
    Caren Garrity, Microwave Group, Dunrobin, Ont.

René Ramseier, Microwave Group, Dunrobin, Ont.
Ken Asmus, Atmospheric Environment Service, Ottawa, Ont.
Bob Writner, Scripps Institution of Oceanography, La Jolla, Calif.

15. **Sea Ice Measurements**
   Terry Tucker, U.S. Army Cold Regions Research and Engineering
   Laboratory, Hanover, N.H.
   Mary Williams, National Research Council, Institute for Marine
   Dynamics, St. John's, Nfld.
   Hazen Bosworth, U.S. Army Cold Regions Research and
   Engineering Laboratory, Hanover, N.H.
   Tony Gow, U.S. Army Cold Regions Research and Engineering
   Laboratory, Hanover, N.H.

16. **Ship Technology**
   Jim St. John, Science and Technology Engineering Corporation,
   Columbia, Md.
   Ron Ritch, A.R. Engineering, Calgary, Alta.
   Larry Schulz, Science and Technology Engineering Corporation,
   Columbia, Md.
   Rubin Sheinberg, U.S. Coast Guard, Naval Engineering Division,
   Baltimore, Md.

17. **Standing Stocks and Production Rates in the Arctic Ocean**
   Pat Wheeler, Oregon State University, Corvallis, Oreg.

18. **Character and Quality of Sediments Transported by Sea Ice**
   Erk Reimnitz, U.S. Geological Survey, Menlo Park, Calif.

19. **Geochemistry of Trace Elements in Arctic Shelf Waters**
   Chris Measures, University of Hawaii, Honolulu, Hawaii

Index

Aagaard, Knut, 8, 10, 14, 28, 29, 58, 71, 99, 105, 117, 124, 136, 153, 154, 172, 174, 192, 195, 198, 208, 209, 216, 220, 230, 242, 244, 246, 248, 252

acid rain, 155, 164, 196

Across the Top of the World (Herbert), 227

Adamson, Louise, 138, 165ff., 212, 219, 229, 230-231, 252

Adushkin, Vitaly, 180

Alabama, 16

Alaska, 19-24, 30, 31, 45, 58, 60-62, 103, 151, 212

albedo, 211, 229, 246

Alert Bay, 87

Aleutian Islands, 30

Almirante Oscar Viel, 85

Alpha-Mendeleyev Ridge system, 70, 71, 124

Alpha Ridge, 188-189, 198, 208, 209, 216

Alps, 150, 151

aluminum, 132, 212

Amagoalik, Ikaksak, 78-80, 102, 106, 148, 156, 212, 237, 253

Amazonian rainforest, 50

American Museum of Natural History, 161

ammonia, 48, 127

amphipods, 139, 203, 213ff., 219

Amundsen, Roald, 9, 24-27, 28, 116, 242

Amundsen Basin, 217

Antarctica, 50, 57, 59, 63, 90, 131, 152, 154, 160, 168, 191

Archer, Colin, 75-76

Arctic, 96-97

Arctic Halocline, 120

Arctic Mid-Ocean Ridge, 70

Arctic Mixed Layer, 120, 186

Arctic Monitoring and Assessment Program, 164

Arctic Ocean, 8, 10, 11, 14, 25, 31, 34, 55, 60, 91, 95, 122, 133, 141, 146, 188, 242, 244; contaminants in, 159-160, 162-164, 185-186; layers in, 115-116, 119-121; rising temperature of, 80, 124-126, 169-170, 173, 217

"The Arctic Ocean and Climate" (Aagaard and Carmack), 117

Arctic region, 38-42, 58-63, 69-72, 86

Arktica, 188, 236

Arkwright, Sir Richard, 51

Arnold, Lt. Karen, 53

Asia, 162, 163, 189

Asmus, Ken, 254

Atlantic Deep Water, 120-121, 134, 244

Atlantic Layer, 120, 124-125, 170, 186, 216-217, 219
Atlantic Mid-Ocean Ridge, 61, 71, 188
Atlantic Ocean, 4, 5, 8, 30, 31, 61, 115, 119, 121, 124, 133, 136, 141, 151, 153, 170, 185, 240, 242, 246
atmosphere of Earth, 49-50
Axel Heiberg Island, 97

Baffin Bay, 94
Baffin Island, 39, 50
Baltic Sea, 160, 163
Banff, Alberta, 111, 145
Bangladesh, 151
Banks, Sir Joseph, 67
Barents Abyssal Plain, 242
Barents Sea, 31, 120, 124, 185, 188, 244, 245, 246
Barometer Rising (MacLennan), 248
Barrow, Sir John, 83
Barwell-Clarke, Janet, 138, 220, 231, 252
Bascom, Willard, 118
Beare's Sound, 41
bears. *See* polar bears
Beaufort Sea, 8, 58, 180
Bedford Institute of Oceanography (BIO), 8, 29, 30, 65, 184, 251, 252, 253
Behaim, Martin, 39
Bennett, James Gordon, 32
Bennett Island, 95
Berger-North, Kent, 64, 68, 72, 251
Bering, Vitus, 30, 83
Bering Sea, 95
Bering Strait, 10, 29, 30, 31, 33, 48, 60, 83, 94, 97-98, 119, 120, 122, 239, 241
Beringia ice bridge, 103
Bernier, Joseph-Elzéar, 9, 94-99
Best, George, 39-41
bioaccumulation, of contaminants, 159-163
Black Current. *See* Japanese Current
bolus, 173, 216
boreal forest, 13
Bosworth, Hazen, 254
bowhead whales, 213

box core, 66, 68, 77, 139, 216, 229
Bradford, William, 74
Brickell, Peter, 251
Brigham, Capt. Lawson, 54-58, 91, 134-135, 181, 191, 195, 201, 202-204, 208, 235, 236-237
brightness temperatures, 147-148, 178
Brooklyn, New York, 151
Brooks Range, 60
Brougham Island, 162
Buchan, Capt. David, 84
burrowing anemones, 67
Byam Martin Island, 97
Byrd, Adm. Richard E., 25, 26

California, 11, 59, 117, 119, 130
Campanian sediment, 189
Canada, 162, 164, 181
Canada Basin, 58, 61, 62, 124-125, 136, 139, 170, 172-173, 174, 186, 188-189, 193, 209
Canadian archipelago, 97-98, 136. *See also* Queen Elizabeth Islands
Canadian Basin, 11, 58-63, 70, 71, 85, 104, 120, 124, 189, 209, 245
Canadian Coast Guard, 10, 44, 53, 85, 86, 89, 146
Canadian Experiment to Study the Alpha Ridge (CESAR), 189
Canadian Maritime Air Command, 206
Canadian Vickers, 87
Canmar Kigoriak, 93
Cape Town, South Africa, 151
carbon, 109, 129, 186
carbon dioxide, 48-50, 124, 128, 131, 133, 138, 167. *See also* greenhouse effect
carbon tetrachloride, 165, 169
Carboniferous period, 51
Carey, William S., 60, 63
Carmack, Eddy, 7, 6-11, 14, 28, 29, 52, 58, 71, 85, 89, 99, 105, 117, 121, 122, 124, 138, 153, 162-163, 170, 172, 174, 175, 186, 191, 195, 208, 219, 220, 228, 229, 244, 249, 252, 253
carousel water sampler. *See* rosette
Carson, Rachel, 170

Cathay, 39, 41, 42
Central Laboratory for the International Study of the Sea, 98
Centre for Research in Experimental Space Science (CRESS), 143
cesium, 181, 183, 185, 244
CFCS, 12, 125, 138, 165ff. *See also* contaminants
Challenger pumps, 77, 139, 229
The Changing Atmosphere (Firor), 14
Charlie Transform Fault, 71-72, 135-136, 139, 174
Charlton Island, 42
Chelyabinsk-40, 182
Chernobyl, 165, 183, 185
Chile, 85, 122
China, 181. *See also* Cathay
Chipp, Lt. Charles W., 34
chlordane, 214
chlorine, 166-168
chlorophyll, 128
Christenson, John, 253
Chukchi Abyssal Plain, 104, 135
Chukchi Borderland, 70, 71, 173
Chukchi Gap, 135, 137, 139, 156
Chukchi Sea, 30-34, 43, 46, 64, 125, 127
Churchill, Manitoba, 102, 108, 163
Churchill, Winston, 144
ciliates, 65
Citadel Hill, Halifax, 248-249
Clark, Joe, 45
climate, 48, 117, 120, 121, 153, 155, 219, 245-246
climate change. *See* global warming
cloud condensation nuclei (CCNs), 196
clouds, 50, 138, 195-196, 214, 219
Clough, Lisa, 66-68, 180, 251
Coachman, Larry, 7, 8
coal, pollution from, 51
Coast Mountains, 60
cobalt, 60, 181
cod, arctic, 159, 163, 164, 213
Coleridge, Samuel Taylor, 42, 43
Columbia River, 122
Columbus, Christopher, 247

Confederation Bridge, 204
contaminants, 110, 138, 159-170, 214, 245. *See also* radionuclides
continental shelf, 101, 104, 127
Cook, Frederick, 80, 221-222, 225
Cook Islands, 171
copepods, 67, 127, 139, 159, 163, 202, 213ff., 219, 245
Cornwallis Island, 78
Corwin, 72-73, 76, 203
The Crest of the Wave (Bascom), 118
Cretaceous period, 51, 61, 189
CTD program, 65, 139, 175, 209, 216, 229
Cusack, Mark, 206
Cytochrome P450, in polar bears, 110

Dallas, Texas, 12
Danish Geographical Journal, 73
Dartmouth, Nova Scotia, 8, 88, 248-249
Darwin, Charles, 34, 129
Darwin's finches, 112
Davis Strait, 31, 67, 97, 180
DDE, 161
DDT, 161, 165, 214
De Long, Lt. George Washington, 32-34, 72, 73, 74, 94, 192
Deep Sea Clamm, 67-68
Defense Meteorological Satellite Program (DMSP), 138
Denmark Strait, 70
density of water, 122
Department of Energy, Mines and Resources, 146
Department of Fisheries and Oceans, 8
Department of Marine and Fisheries, 96
Des Grosseillers, 85
desertification, 155, 195
dimethylsulfide (DMS), 214
dimethylsulfoniopropionate (DMSP), 214
Diomedes Island, 30
dirigibles, 25, 27
Dome Petroleum, 93
Dorothea, 84
Drake, Leslie, 223ff.
drift stations, 62, 119, 173, 188

du Pont, 165, 168
dust, iron in, 128-134, 212

Early Voyages and Northern Approaches
    (Oleson), 38
East Siberian Sea, 33, 34, 73, 188
Ebbysmeyer, Curt, 123
Eglinton Island, 97
Einstein, Albert, 14, 121
Ekwurzel, Brenda, 29, 139, 184-186, 207,
    244, 253
El Niño, 5, 122
Ellef Ringnes Island, 97, 137
Ellesmere Island, 81, 87, 124
Elliott, Jim, 29, 30, 84, 252
Ellis, Kathy, 184, 244, 253
Ellsberg, Edward, 34
Ellsworth, Lincoln, 25, 26
The End of Nature (McKibben), 164
England, 41, 153, 181, 185
Environment Canada, 8
Environmental Protection Agency (EPA),
    162
ERS-1 satellite, 194
Eurasian Basin, 62, 120, 124-125, 245
evaporation, 50. *See also* desertification
extinctions, 50-51, 122

Falcon Islands, 39
Farley, Sean, 28, 101, 105, 108, 109, 134,
    138, 253
Farthest North (Nansen), 95, 99, 119
Fedorov, Ivan, 30
Finland, 164
Firor, John, 14, 167
fisheries, disappearance of, 8
flagellates, 65
floods in U.S. (1993), 2-5, 16-17
Florida, 16, 195, 203
fluid dynamics, 14, 121-122
fluorometer, 65
formaldehyde, 130
Forsythe, Louise, 22
Fort McMurdo, Antarctica, 90
Fortunate Isles, 40

Forum on Global Climate Change
    (1989), 12, 13
Fourier, Jean-Baptiste-Joseph, 49
Fram, 69, 75-76, 95-96, 97, 116, 117
Fram Strait, 61, 120, 124, 178, 230, 241, 244
France, 31, 181, 185
Franklin, Sir John, 32, 84, 99
Franz Josef Land, 76, 77
Freon, 125, 165ff. *See also* CFCs; contami-
    nants
Frobisher, Martin, 39-41
Frobisher Bay, 67
Funk Island, Newfoundland, 110

Galápagos Islands, 129, 131
Gander, Newfoundland, 143, 146
Garrity, Caren, 5, 29, 105, 137-138, 141-142,
    172, 174, 178, 206-207, 210-211, 215, 253
Gauss, 96, 97
Geen, Charles, 253
Geographica Universalis, 38
Geology of North America, 58
George's Island, 248
Georgia, 16-17
Germany, 150
glaciation, 48. *See also* ice ages
glaciers, 150-152, 248
global warming, 10, 38, 47-48, 195-197,
    211; in Arctic, 80, 81, 110, 117, 152-154,
    169-170, 178, 185, 245-246; effects of, 5,
    11-13, 48-52, 149-154, 163, 178
Global Warming (Schneider), 14, 152
Glover, Danny, 102
Goethe, Johann Wolfgang von, 150
Gosselin, Michel, 202, 213, 243, 251
Gow, Tony, 195, 254
GPS (Global Positioning System), 69, 80,
    194, 198, 200, 201, 202, 223ff.
Grand Banks, 122, 173
Grandy, Capt. Phil, 28, 29, 35, 43-46, 56,
    77, 87, 90, 99, 114, 143, 159, 172, 174, 175,
    191, 193, 195, 205-206, 208, 212, 220,
    223ff., 231, 236, 240
Grantz, Art, 58-63, 66, 69-72, 135-138, 139,
    172, 187ff., 192, 198, 207, 216, 242-243, 252

gravimeter, 91

Great Gull Island, 160-161

Great Lakes, 13, 52, 122, 152, 160, 204

greenhouse effect, 12, 13, 14, 48-52, 170.
See also carbon dioxide; global warming; methane

Greenland, 33, 38, 39, 45, 48, 49, 61, 70, 73, 75, 95, 103, 124, 133, 144, 150, 154, 164, 189, 195-196, 240

Greenland Ice Sheet, 151

Greenland Sea, 5, 8, 10, 60, 73, 75, 84, 105, 118, 133, 148, 154, 180, 188, 220, 240, 244

Grise Fjord, 78, 106-107

grizzly bears, compared with polar bears, 101-102

Grovenstein, John, 195-196, 252

Gulf Stream, 31, 120, 121, 153-154, 173, 185, 242

gulls: Bonaparte's (*Larus philadelphia*), 84; glaucous (*L. hyperboreus*), 19, 21, 30; greater black-backed (*L. marinus*), 248; herring (*L. argentatus*), 20, 248; ring-billed (*L. delawarensis*), 240; Sabine's (*Xema sabini*), 84; western (*L. occidentalis*), 20

Gvozdev, Mikhail, 30

Halifax, Nova Scotia, 87, 88, 176, 248-249

Halifax-Dartmouth Industries Limited, 88

Hall, Charles Francis, 94,

halocline, in Arctic Ocean, 217

Hansen, James, 12

Hart, Pat, 199, 252

Hawaii, 189

Hays, Helen, 160-161

Healthsat A satellite, 115, 194

Heisenberg, Werner, 14, 219, 221

helium, 186

Helland-Hansen, Bjorne, 117-118, 122

Helsinki Accord, 169

Hemeon, Mike, 105, 174, 207, 223ff.

Hemphill, Steve, 35, 36, 105, 108, 112, 174, 207, 231-232, 240-241

Henrietta Island, 34

Henry Larsen, 85-86, 125, 170, 173

Henson, Matthew, 221, 225

Herald Island, 33, 192

Herbert, Wally, 80-81, 222, 226-227

Hingston, Mike, 138, 165, 253

HNLCs (high nutrient/low chlorophyll areas), 129-133

Høeg, Peter, 94

Holocene epoch, 197

hot spots, 189

House of Commons Standing Committee on the Environment (1989), 12, 13

hovercraft, as icebreakers, 146

Hudson Bay, 97, 102

Humboldt Current, 5, 121, 122

hydrogen, 186

IBM, 168

Ibn Sa'id, Abu'l-Hasan 'Ali, 39

ice: age of, 43, 210; anchor ice, 177; in Antarctic, 63; as building material, 144-145; as living organism, 204-205; dirty ice, 74, 177-181; drift patterns of, 177ff.; glacial, 149-150, 151-152, 197; pack ice, 144; properties of, 37, 41-42, 43, 81-82

ice ages, 48, 50, 133, 149-151, 197, 248

ice algae (*Melosira arctica*), 203, 214, 243

ice blink, 69

ice cap, 133, 152, 191

Ice Centre, 36, 87, 141

ice islands, 40, 119, 185, 189

ice maps, 36, 142ff. See also remote sensing

ice observers, 134, 141, 206

Ice Patrol, 87, 141-142, 146

ice stations. See drift stations

Ice Time (Levenson), 14

icebergs, 43, 113, 141

icebreakers, 85, 89, 130, 188, 204, 214-215, 236; design of, 91-94; nuclear-powered, 87, 90, 181; Russian, 56, 85, 89, 181, 223

Iceland, 70, 164, 189, 242-243, 244

Iceland Sea, 121

Iceman, 102

Idaho National Engineering Laboratory, 165

Illinois, 4

Indian Ocean, 153

Indonesia, 168

Industrial Revolution, 51, 52

infrared line scanner, 146

Inland Waters Directorate, 146

Inmarsat satellites, 35, 36, 115, 142, 194

Institute of Ocean Sciences (IOS), 6, 8, 9, 28, 85, 155, 165, 173, 251, 252, 253

International Glaciological Society, 195

Inuit, 22, 33, 48, 72, 78, 97, 102, 103, 148, 163, 205, 221, 225

iodine, 185

Iowa, 3, 4

Iqaluk, Oolatita, 103, 105-107, 148, 156, 163-164, 209-210, 212, 219, 247, 253

Irish Sea, 185

iron, 128-134, 158, 203, 229

IRONEX-93, 131

Italia, 27

Izvestiya, 182

Jackson, Frederick, 77

jaegers: pomarine (*Stercorarius pomarinus*), 78, 83-84

James, Capt. Thomas, 42

James Bay, 42

Jan Mayen Fracture Zone, 71

Jantunen, Liisa, 139, 252

Japanese Current, 31, 33

Jeannette, 32-34, 72-76, 93, 95, 178, 192, 203, 234

Jeannette Island, 34

Jensen, Sören, 160

Jet Propulsion Lab, 147

Jewison, Norman, 102

Johansen, Frederik Hjalmar, 76

John A. Macdonald, 45, 85, 89

Johns, Capt. David, 10, 14, 45, 85, 86, 87, 89, 91, 191

Jones, Peter, 29, 189, 235, 237, 251

Josephine Ford, 25

Jurassic period, 51

Kane, Elisha Kent, 99

Kara Sea, 179, 180, 181ff.

Karachai, Lake, 183

Kellett, Henry, 32

King Christian Island, 97

Kipling, Rudyard, 15, 16

kittiwakes: black-legged (*Rissa tridactyla*), 29, 78, 83-84, 164, 246, 248; red-legged (*R. brevirostris*), 29

Kodz, Fred, 134, 172, 174, 207

komatiks, 80, 107, 155-156

Krasilov, Gennady, 180

Kula tectonic plate, 60

Kuro-Si-Wo. *See* Japanese Current

La Guardia Airport, 151

La Jolla, California, 8, 117

Labrador, 153

Labrador, 44

Labrador Current, 5, 122, 141

Lafford, Tom, 21

Lambert, Gustav, 31

Land Evaluation Group study (1989), 13

Laptev Sea, 61, 71, 124, 179-180, 182, 240, 244

Laurier, Sir Wilfrid, 96

leads, in sea ice, 77, 81, 104, 134

Lee, Alethia, 206

Lena River, 34, 179, 244

Lenin, 181

Lenoir, Jean Joseph Étienne, 51

Levenson, Thomas, 14

LIMEX (Labrador Ice Margin Experiments), 143

Lincoln Experimental Satellite (LES-9), 194-195

Little Ice Age, 48

Littleton Island, 94

Livingstone, David, 32, 77

Lomonosov Ridge, 61-62, 124, 125, 188, 192, 198, 208-209, 213, 215-216, 229

Long Island Sound, 160

Lopez, Barry, 63, 90, 189

Los Angeles, California, 151

Louis S. St. Laurent, 16, 28, 35, 43, 46, 53, 56, 61, 77, 84, 92-94, 101, 117, 130, 136, 139, 159, 173, 184, 193, 239, 244; midlife refit, 87-90; reaches North Pole, 219-237
Louise, Lake, 145
Lubin, Dan, 252

Maastrichtian sediment, 189
Macdonald, Robie, 28, 125, 139, 159-160, 163-164, 186, 213, 252
Mace, Bonnie, 251
Mackenzie, C. J., 144
Mackenzie Delta, 62, 70
MacLennan, Hugh, 248
Makarov Basin, 124, 125, 170, 185, 191, 202, 209, 217
Manhattan, 44, 89
Mare Island Navy Yard, 32
Marisat, 194
Martha L. Black, 86
Martin, John, 129, 131
May, Steve, 199, 252
Mayak Nuclear Facility, 182
M'Clure Sound, 89, 90
McKibben, Bill, 164
McLaughlin, Fiona, 28, 125-126, 138, 165ff., 186, 220, 230, 252, 253
McNab's Island, 248
Measures, Chris, 28, 128-134, 139, 158, 203, 212, 254
Mediterranean Sea, 38
melt ponds, 134, 149, 154, 178ff., 211, 243
Melville, Herman, 113
Melville Island, 97
Mendeleyev Abyssal Plain, 136
Mendeleyev Ridge, 125, 134, 172, 174, 188, 191, 209. *See also* Alpha-Mendeleyev Ridge system
methane, 48, 124
Mexico, Gulf of, 4, 5, 203
Microwave Group, 29, 143, 253, 254
Minnesota, 3, 4
Mirsky, Jeannette, 221
Mississippi River, 3, 4, 11
Mississippi Valley, 4, 5, 16

Missouri River, 2, 3, 4, 11
Mohn, Henrik, 73
Molina, Mario, 167
Monsanto, 161-162
Montreal Protocol on Substances That Deplete the Ozone Layer, 168-169
Moodie, Superintendent, 97
Morse code, 114, 194
Moscow News, 183
Mountbatten, Lord, 145
Mowat, Farley, 222
Muir, John, 72, 203
Murmansk, Russia, 233, 239, 241
Muus, David, 252
My Life as an Adventurer (Amundsen), 26

Nansen, Fridtjof, 73-77, 94, 95, 97-98, 99, 116-118, 119, 120, 180, 221
Nansen Basin, 217
Nansen bottles, 116
Nares Strait, 246
NASA-Goddard Space Institute, 12, 154, 168
Nathaniel B. Palmer, 63, 93, 189
National Oceans and Atmospheric Administration (NOAA), 137-138, 194
National Research Council (Canada), 144-145, 254
National Research Council (U.S.), 162
Natural History magazine, 161
Natural Resources Defense Council, 162
Nature magazine, 168
Nautilus, 57, 62
Nebraska, 4
Nelson, Rick, 184, 253
Netherlands, 151
New Siberian Islands, 61, 75-76
New York City, 11, 151
New York Herald, 32, 34, 72, 77, 83
Newcomb, Raymond Lee, 32
Newfoundland, 43, 44, 110, 143
Newton, Mount, 7
Niskin bottles, 65, 116, 118, 130
Nitoslawski, Stefan, 21, 35, 53, 65, 77, 176, 193, 220, 233ff.

nitrates, 127, 129, 131
nitrogen, 169-170
nitrogen dioxide, 166
nitrogen monoxide, 166
Nobile, Umberto, 25-27
Nome, Alaska, 16, 19-24, 29, 47, 242, 248
The Noose of Laurels (Herbert), 222, 226
Norge, 25-27, 116
Norman McLeod Rogers, 85
North Carolina, 196
North Dakota, 3, 4
North Polar Basin, 97, 117
North Pole, 9, 10, 11, 14, 57-58, 61, 79, 99, 103, 104, 119, 124, 154, 208, 215; AOS '94 reaches, 219-220, 223-224, 227-237; difficulty of locating, 171-172, 194, 225-227, 228-229; historic trips to, 25-27, 31-34, 40, 57, 73-77, 84, 95-97, 188, 221-223, 224-227; ozone hole at, 169; sovereignty at, 45-46, 99; warming at, 52, 152; wedding at, 212, 220, 230-231
North Water Polynya, 180
Northcliffe, Lord, 77
Northern Sea Route, 55-56
Northumberland Strait, 204
Northwest Passage, 24, 32-33, 42, 44, 86, 89, 239, 240
Northwest Territories, 62, 212
Northwind, 45, 183-184, 203
Northwind Ridge, 62, 70, 71, 188
Norway, 25, 120, 162, 164, 184, 220, 240, 244
Norwegian Directorate for Nature Management, Report of (1994), 181-182
Norwegian Sea, 118
Nova Scotia, 150, 173
Novaya Zemlya, 119, 180, 183
nuclear bomb tests, 129, 180-181, 184
nuclear waste, 165, 180-186
nutrients, in sea water, 127, 131, 214
Nye, Captain William, 74-75

Ob River, 182
O'Brien, Mary, 251
oceanography, 7, 115-116, 118

Oden, 9, 125, 176
Ohio River, 11
oil industry, 11, 58, 93
Oleson, Tryggvi J., 38-39
Olson, Walt, 252
Ontario, 12, 13, 15, 150, 156, 242, 244
Ontario, Lake, 146, 240
Open Polar Sea theory, 31-33, 40, 74, 99, 180
organochlorines, 138, 157, 159-160, 162
Orkney Islands, 38
Osborne, Elizabeth, 252
Oslo, Norway, 56, 75
O'Toole, Kevin, 252
Ottawa Citizen, 89
oxygen, 50, 67, 166ff., 185-186
ozone layer, 124, 165ff.

Pacific Ocean, 4, 5, 8, 31, 33, 59, 60, 115, 120, 122, 129, 131, 151, 153, 246
Panama, Gulf of, 129
passive microwave radiometer, 147-149, 211. *See also* remote sensing
Paton, Dave, 155, 176, 252
Payne, Fred, 199, 253
PCBS, 110-111, 160-164, 165, 214. *See also* contaminants
Pearson, Rick, 252
Peary, Robert Edwin, 80, 221-223, 224-226
peregrine falcons, 162
Perkin, Ron, 125, 175, 209, 252
pesticides, 161-162
Petermann, August, 31, 32, 33, 61-62
Peterson, D. J., 182
Petrovich, Ferdinand. *See* Wrangel, Baron von
Phillips, Larry, 253
phosphates, 129-132
photosynthesis, 129, 164
phytoplankton, 65, 119, 127-134, 163, 167, 184, 214, 245
Pierre Radisson, 85
Pinatubo, Mount, 169
piston core, 136, 137, 139, 174, 216
Plaisted, Ralph, 222

plate tectonics, 59, 62
plutonium, 180, 184, 185
Point Barrow, Alaska, 45, 46, 86
Polar 8, 86-87, 88, 89
polar bears (*Ursus maritimus*), 28, 35, 39, 101-112, 157-158, 213, 219, 245; contaminants in, 110-112, 162-163, 214
The Polar Passion (Mowat), 222
Polar Sea, 16, 28, 45, 46, 53-58, 61, 63-64, 77, 86, 90-91, 101, 117, 136, 139, 173, 180, 190, 202ff.; propeller problem, 215, 220, 229-230, 239, 241; reaches North Pole, 219-237
Polar Star, 57, 70, 86, 215
Polaris, 94
Polarstern, 5, 9, 105, 142, 143, 148, 215, 242
Pole of Relative Inaccessibility, 191
polynyas, 104, 180, 202, 244
polyvinyls, 160
Powers, Mike, 84, 192
pressure ridges, 43, 63, 79, 81, 104, 135, 155, 174, 227
Pribilof Islands, 30
primary production, in sea water, 127-134, 184
Prince Edward Island, 204
Prince Patrick Island, 97
Prince of Wales, Cape, 30
Prince of Wales Strait, 89
Project Habbakuk, 144-146
puffins: tufted (*Fratercula cirrhata*), 30; horned (*F. corniculata*), 30
Pyke, Geoffrey, 145
Pykerete, 145
Pytheas, 38

quantum physics, 14
Quebec, 160
Queen Elizabeth Islands, 45, 162. *See also* Canadian archipelago

Radarsat, 211
radiation, solar, 48, 153-154, 178, 246
radioactivity, 129
radionuclides, 164, 165, 180, 184-185, 219

radium, 184
rafting, 43, 157
Raleigh, North Carolina, 196
Raleigh, Sir Walter, 41, 118
Ramsay, Malcolm, 28, 101-106, 128, 134, 138, 159, 163, 190, 213-214, 247, 249, 253
Ramseier, René, 143-144, 206-207, 211, 254
Raycon beacon, 157
Reagan administration, 11, 211
reflectivity, of ice, 146-149, 178ff. *See also* albedo
Reimnitz, Erk, 177, 179, 188, 202, 243-244, 254
remote sensing, 137, 141-149, 193-194, 211-212, 219
Resolute, 78, 114, 115
Rime of the Ancient Mariner (Coleridge), 42
Risebrough, Robert, 160-161
Ritch, Ron, 92-94, 175-176, 254
Robinson, Bill, 199, 253
Rocky Mountains, 59, 60, 62, 172
rosette, 64-65, 77, 117, 130, 139, 173, 209, 229, 242
Ross, John, 67
Ross Ice Shelf, 57
Rowland, F. Sherwood, 167
Royal Air Force, 144
Royal North West Mounted Police, 97
Russia, 30, 31, 46, 102, 109, 119, 164, 177, 181, 232

Saanich Indians, 6
Sabine, Edward, 84
Sabine, Joseph, 84
salinity of water, 65, 81, 115, 119, 120, 122, 138, 153, 185, 209
San Francisco, California, 32, 74
Sannikof Island, 95
SAR (synthetic aperture radar), 87, 147
Saskatchewan, 13
satellite imagery. *See* remote sensing
Schmitt, Jim, 138, 253
Schneider, Stephen, 12, 14, 152
Schulz, Larry, 254

Scientific American magazine, 152
Scott, Frederick, 25
Scripps Institution of Oceanography, 8, 29, 65, 117, 118
Sea of Japan, 181
sea level rise, 149-154
sea mount, discovery of, 209-210
sea snails, 67
seals: contaminants in, 162, 163; ringed (*Phoca hispida*), 103, 104, 106-107, 109, 113
Seasat, 147
Seastar pumps, 139, 156-159, 164, 174, 229
Seattle, Washington, 239, 242
Section Bay, 146
seismic tow, 135, 139, 172, 173ff., 176, 198, 208, 216
Sellafields nuclear facility, 185, 244
Senate Committee on Energy and Natural Resources (1988), 12, 13
Seward Peninsula, Alaska, 20, 29
Sheinberg, Rubin, 254
Sherr, Evelyn, 197, 251
Siberia, 30, 31, 32, 33, 50, 60, 61, 62, 71, 74, 95, 101, 120, 175, 182ff., 188, 244
Siberian Abyssal Plain, 202
Sidney, B.C., 6
Sieberg, Doug, 204, 212, 219, 229, 230-231, 253
Simard, Natalie, 251
SLAR (side-looking airborne radar), 147
Smilla's Sense of Snow (Høeg), 94
Smirnoff, Captain, 231, 236
snowmobiles, 79-80, 81, 107
solar radiation, 48, 153-154, 178, 246
South Africa, 151
South Dakota, 4
South Pole, 25
sovereignty, in Arctic, 44-46, 78, 87, 96, 97, 99
The Soviet Maritime Arctic (Brigham), 55
Soviet Union, nuclear program, 180-186
Spitsbergen, 24, 25, 26, 31, 61, 84
Stanley, Henry Morton, 32, 77
St. John, Jim, 92-94, 194, 254

St. Lawrence, Gulf of, 85, 86, 87, 122
St. Lawrence River, 122, 151
St. Michael's, Alaska, 33
State of the Environment Report for Canada, 169
The State of Habitat Protection in the Arctic, 181
Steller, Georg Wilhelm, 83
Stoodley, Gord, 35, 36, 113-114, 194
stratosphere, 166, 169
Strello, Mount, 73
strontium, 181, 183, 185
SSM/I satellite, 193
Svalbard, Norway, 28, 111, 162, 163, 214, 240
Sverdrup, Harold, 117
Sverdrup, Otto, 73, 97, 116, 117
Svino Chasm, 60
Sweden, PCBs in, 160
Swift, Jim, 29, 117-118, 126, 139, 153, 172, 173-174, 216, 252
Switzerland, 150

Techa River, 182-183
Telazol, 108
Teller, Alaska, 24, 26, 27
temperature of water, 122, 246; in Arctic Ocean, 116, 119, 120, 170, 173, 209
TeraScan, 137, 141-142, 172, 193-194, 201, 206
terns, 160-161, 163
Thailand, 151
thermal expansion, 152
thermal radiation, 48
Thibault, Delphine, 67, 72, 251
thorium, 184
Thule, Greenland, 45, 144
Titanic, 141, 230
To the Arctic! (Mirsky), 221
To the North! (Mirsky), 221
Tokyo, Japan, 56, 151
Tom River, 182
Tomsk-7, 182-183
Transport Canada, 89
Trent, 84

Triassic period, 51
tritium, 139, 186
Tromsø, Norway, 31, 239, 242
Tropical Storm Alberto, 16
troposphere, 161
Troubled Lands (Peterson), 182
Tucker, Terry, 177-179, 188, 195, 202, 243, 254
Tuele, Darren, 155, 176, 252
Tuktoyaktuk, 188, 242

U-boats, 144-145
ultraviolet radiation, 166-167
Ungava Peninsula, 146
United States: nuclear waste from, 164, 165, 181; PCBs banned in, 162
upwelling, 132
Ural Mountains, 183
uranium, 165, 184
U.S. Air Force, 195
U.S. Atomic Energy Commission, 129
U.S. Coast Guard, 27, 53-57, 131, 141, 183, 195, 203-204, 254
U.S. Food and Drug Administration, 162
U.S. Meteorological Office, 5
U.S. Polar Research Board, 9

Vancouver, B.C., 86, 97, 151
Vancouver Island, 59
Victoria, B.C., 9, 15-16
Viking, 73, 75
Vikings, 38-39, 48, 197
Vinland, 38
Viscount Melville Sound, 44
volcanos, 167

Walsh, Michael Joseph, 22
Warwick Island, 41
Washington, D.C., 11
Washington State, 59
Watt, James, 51
Weddell Sea, 63
Wegener, Alfred, 60
Whales, Bay of, 57
Wheeler, Pat, 127, 203, 251, 254
Wheeler, Steve, 192, 200, 201, 208
Williams, Bob, 253
Williams, Mary, 79, 80-82, 175, 195, 204-205, 212, 214, 254
Wisconsin, 3, 4
Wisconsin Ice Age, 50, 149, 151
Woods Hole Oceanographic Institution, 118
World Ocean Circulation Experiment (WOCE), 13
worms, 66, 67
Wrangel, Baron von, 32, 83
Wrangel Island, 72, 74, 109
Wrangel Land, 32, 33, 61, 72
Writner, Bob, 193, 254

Yablokov, Aleksei, 181
Yamal, 223, 224, 231-237, 239, 241
Yenisei River, 179
Yukon River, 33
Yukon Territory, 244

Zemlyak, Frank, 138, 165, 176, 219, 235-236, 237, 252
zooplankton, 65-66, 127-128, 167, 213ff., 229